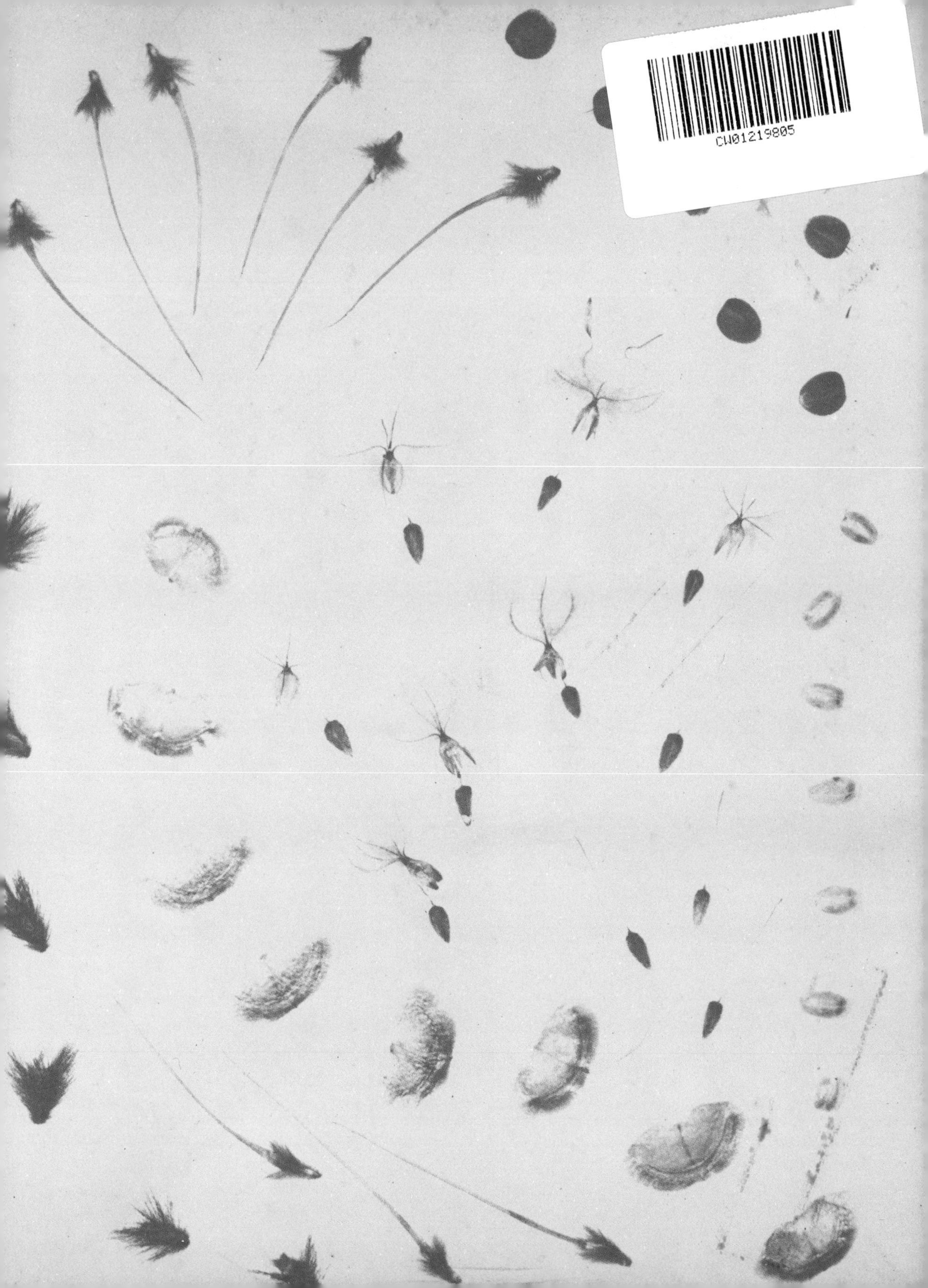

Proteas for Pleasure

Other books by the same author:

Wild Flowers of Southern Africa

The Complete Gardening Book

Discovering Wild Flowers in Southern Africa

Bulbs for the Gardener in the Southern Hemisphere

Gardening the Japanese Way

Namaqualand in Flower

Shrubs, Trees and Climbers for Southern Africa

Garden Beauty of South Africa

Leucadendron argenteum, Protea compacta, Protea magnifica (pink and yellow), *Protea cynaroides, Mimetes cucullatus, Leucadendron tinctum, Berzelia lanuginosa* (not a member of the Protea family) and *Protea sulphurea* (lying on the table).

PROTEAS
FOR PLEASURE

How to grow and identify them

SIMA ELIOVSON

M

This book is copyright under the Berne Convention. No portion may be reproduced by any process without written permission. Inquiries should be made to the Publisher.

Published by
MACMILLAN SOUTH AFRICA PUBLISHERS (PTY) LTD
Braamfontein Centre,
Jorissen Street,
JOHANNESBURG
Associated companies
throughout the world

1st Edition, 1965
2nd Edition, 1967
3rd Edition, Enlarged and Revised, 1973
4th Edition, 1976
5th Edition, 1979
6th Edition, Revised, 1983

ISBN 0 86954 006 8

Printed and bound in Hong Kong

LURE OF THE PROTEA

Where silver rocks reflect the mountain streams,
The Protea holds aloft its stately head—
Majestic flower to prompt the thoughts of men,
According to their nature and their dreaming.

To some it seems a giant artichoke,
While others link it to the ancient god
for whom 'tis named—
The ruby glow and allied silky beard
evoke the mystery of primeval days;
A traveller, who ranges far from home,
will quicken at its sight and yearn
And envy sugarbirds,
that dip into the gleaming cup,
while swaying gently in the wind.

SIMA ELIOVSON

FOREWORD

PROFESSOR H. B. RYCROFT

Director, National Botanic Gardens of South Africa, Kirstenbosch, Cape Town.

Through the publication of her other books Sima Eliovson has rendered signal service in enlightening her readers about the beauty and uniqueness of our indigenous plants. She has also made us realise the importance of preserving these gifts of nature and has encouraged us to grow them in our private gardens.

I am particularly happy to introduce her latest book to the ever-increasing numbers of people who are taking more and more interest in the natural flora of South Africa and especially in the fascinating Protea family.

In this book she has presented a wealth of detail about one special group of our flora and has given the information about many questions to which we would like to have the answers. While great pains have been taken in the preparation of this book to ensure scientific accuracy, it has been attractively written in non-technical language to make it interesting for the layman. In fact, Mrs. Eliovson has succeeded in achieving a happy combination of the scientific and the aesthetic.

The copious illustrations are of excellent quality and should make it easy for anyone to recognise and identify the relevant species. The sections on cultivation and horticultural requirements are certain to help and inspire every Protea gardener.

We are indeed most grateful to Sima Eliovson for writing about the Protea family in such a readable, interesting and informative manner.

H. BRIAN RYCROFT

CONTENTS

Foreword vii
List of Colour Illustrations xi
Acknowledgements xv
Protea—Flower of a Nation xviii
Introduction xix

PART I

1. PROTEAS AND THEIR NAMES 3
   Scientific and Popular Names – The Protea Family – Plant Geography of the Protea Family – Maps showing Place Names – Members of the Protea Family throughout the World – Popular Questions about Proteas.

2. ECHOES OF THE PAST 11

3. PROTEAS RARE AND COMMON 16

4. ON GROWING PROTEAS 18
   The Soil – How to Correct Poor Drainage – How to Prepare the Planting Hole in Normal Soil – Acidity and Alkalinity of the Soil – Watering Proteas – Mulching – Situation – Protection Against Frost – Pests and Diseases – Care after Flowering – Pruning – Growing Proteas in a Nutshell.

5. CULTIVATION OF PROTEAS AROUND THE WORLD 31

6. PROPAGATION OF PROTEAS 33
   Growing Proteas from Seed – Types of Seeds – Fertile Seed – The Time to Sow Seed of Proteas – How to Prepare the Soil – Transplanting into the Open Ground – Growing Proteas from Cuttings.

7. CHOOSING PROTEAS 41
   The Pick of Proteas – First Choice – Second Choice – Classification of Colours in the Protea Family – Proteas throughout the Year.

8. LANDSCAPE DESIGN WITH PROTEAS   48
   Larger Types of Proteas – Medium-sized Proteas – Types of Foliage – Low or Spreading Proteas for Rockeries – Proteas for Pots or Tubs – How to Plant a Protea Collection.

9. PROTEAS AS CUT FLOWERS   57
   On picking Proteas.

PART II

Individual Descriptions of Members of the Protea Family, with Illustrations   63

PROTEAS AS A COMMERCIAL CUT FLOWER CROP   222

Photographic Acknowledgements   224
Index   226

MEASUREMENTS

For the convenience of gardeners, the following conversion table will cover any measurements referred to in this book.

Inches to Centimetres		*Feet to Metres*	
½″	1,27 cm	1′	.305
1″	2,54 cm	2′	.610
1½″	3,81 cm	3′	.914
2″	5,08 cm	4′	1,219
3″	7,62 cm	5′	1,524
4″	10,16 cm	6′	1,829
5″	12,70 cm	7′	2,134
6″	15,24 cm	8′	2,438
7″	17,24 cm	9′	2,743
8″	20,32 cm	10′	3,048
9″	22,86 cm	11′	3,353
10″	25,40 cm	12′	3,658
11″	27,94 cm	15′	4,572
12″	30,58 cm	20′	6,096
15″	38,10 cm	25′	7,620
18″	45,72	30′	9,144
20″	50,80 cm	1000′	±300
24″	60,96 cm		
36″	91,44 cm		

LIST OF COLOUR ILLUSTRATIONS

Frontispiece: *Protea magnifica, P. compacta, P. cynaroides, P. sulphurea, Mimetes cucullatus, Leucadendron tinctum, L. argenteum, Berzelia lanuginosa* (not a member of the Protea family)

Aulax umbellata	207
Brabejum stellatifolium	205
Diastella divaricata	211
Faurea saligna	213
Leucadendron album	187
L. arcuatum	177
L. argenteum	175
L. coniferum	183
L. discolor	181
L. eucalyptifolium	181
L. galpinii	189
L. globosum	175
L. linifolium	189
L. muirii	183
L. platyspermum	185
L. rubrum	185
L. salignum	179
L. sessile	177
L. uliginosum	187
L. xanthoconus	179
Leucospermum bolusii	173
L. catherinae	155
L. cordifolium	147
L. conocarpodendron	151
L. gerrardii	161
L. glabrum	165
L. grandiflorum	169
L. lineare	163
L. mundii	171
L. oleifolium	171
L. praecox	157
L. prostratum	159
L. reflexum	149

L. rodolentum	173
L. tottum	153
L. truncatum	167
L. vestitum	165
Mimetes argenteus	193
M. cucullatus	191
M. hirtus	195
Orothamnus zeyheri	209
Paranomus reflexus	203
P. spicatus	203
Protea acaulos	133
P. acuminata	95
P. amplexicaulis	101
P. aristata	91
P. aspera	145
P. aurea	105
P. burchellii	119
P. caffra	83
P. compacta	109
P. cordata	141
P. coronata	129
P. cryophila	143
P. cynaroides	65
P. effusa	93
P. eximia	115
P. grandiceps	77
P. humiflora	139
P. hybrid	89
P. lacticolor	121
P. laetans	87
P. laurifolia	81
P. lepidocarpodendron	111
P. longifolia	107
P. magnifica	67
P. mundii	123
P. nana	127
P. neriifolia	79
P. nitida	73
P. obtusifolia	71
P. pendula	97
P. pityphylla	93
P. pudens	125
P. repens	69

P. restionifolia	145
P. roupelliae	85
P. rubropilosa	87
P. scolopendriifolia	139
P. scolymocephala	131
P. speciosa	75
P. stokoei	117
P. subulifolia	103
P. sulphurea	99
P. susannae	113
P. venusta	137
P. witzenbergiana	135
Serruria aemula	201
S. barbigera	199
S. burmanii	199
S. florida	197
S. hybrid and parents	201
S. pendunculata	201
Spatalla curvifolia	211
Telopea speciosissima (Waratah)	217
Vegetative reproduction	89

Cuttings by mist-spray
Air-layering

BLACK AND WHITE ILLUSTRATIONS

Sugarbird, Cape	219
Sunbird, Orangebreasted	220
Protea seeds	221

xiii

ACKNOWLEDGEMENTS

Much of the pleasure I have had in preparing this book has been in sharing the project with fellow enthusiasts and friends who have tried to help me in every way. I have enjoyed the meetings with growers, collectors and mountaineers, whose observations and experiences have served to implement my own, and felt encouraged by their enthusiasm when the task sometimes seemed too ambitious.

I wish to acknowledge my thanks to Dr. L. E. W. Codd, Chief of the Botanical Research Institute, Pretoria, who has put the Facilities of the National Herbarium at my disposal, and to Mrs. L. du Toit and Miss H. Gerber, who have helped in checking data and scientific names. I am also indebted to Dr. Margaret Levyns of the University of Cape Town for information on *Paranomus* and to Ion Williams for help in identifying specimens of *Leucadendron*.

My grateful thanks are extended to Professor H. B. Rycroft for kindly agreeing to write the foreword, as well as for information on some of the rarer Proteas.

I owe a debt of sincere gratitude to Mrs. Ruth Middelmann, who sent me many fresh flowers by air from the Cape, which I was able to photograph, as well as to Dr. P. J. Joubert, whose indefatigable enthusiasm sent him into the Cape mountains to collect flowers for photographing and whose observations were extremely helpful. My thanks are also due to Dr. David Dodds, who brought me *Protea rubropilosa* flowers from the Wolkberg.

The photographs were selected from many hundreds. The flowers were sometimes re-photographed several times in the effort to capture the essential character, as well as the beauty of each type. This was particularly important in the black-and-white pictures, where it was necessary to recognise the different species without the aid of colour. The concept of this book began several years ago and at least half of the photographs were taken by my late husband, Ezra Eliovson. The remainder were taken by myself on a Hasselblatt camera, once I had learnt to operate it.

The variety and rarity of species in the Protea family and the necessity to produce this book in time for my ever-patient publisher, led me to ask several people for their assistance, which they gave generously. Professor H. B. Rycroft kindly sent me his negatives from which to make my choice, and I have used five of his pictures. Dr. L. E. W. Codd allowed me to use two pictures of *Faurea* from the Botanical Research Institute, which I superimposed upon one another. Stuart Macpherson lent me two slides of rare plants that he had climbed the mountains to photograph, while Dr. P. P. du Toit sent me a lovely slide of *Protea pityphylla*, when he heard that I was dissatisfied with my own. I appreciate their warm generosity more than I can express.

I should like also to thank Connie Scrimgeour, who printed my black-and-white negatives so beautifully and obligingly. This was no mere professional assignment, but one of friendship and the wish to make the pictures as perfect as possible. To Vic Jacobs, too, there is appreciation for his patience in photographing, several times, my pattern of seeds for the end papers. My gratitude is due to Professor C. J. Uys, who photographed the Cape Sugarbird especially for this book.

My mind is filled with the names of nature-lovers, both at home and abroad, who have added their contribution to my experience. While they are too numerous to mention, my appreciation of their friendly interest is profound.

SIMA ELIOVSON

First edition, 1965. 16, North Road
Dunkeld West,
Johannesburg.

PREFACE

Since the first writing of this text in 1965, the layman's interest in the family *Proteaceae* has increased considerably. Amateur botanists have been fascinated in trying to identify species that occur in the wild and have been baffled by some of the problems that arise and mystify all but the most erudite botanists and authorities on the different genera in this large family.

Revisions of two important genera, *Leucospermum* and *Leucadendron*, have been made by Dr. J.P. Rourke and Dr. Ion Williams respectively, and their monographs have been published during 1972. I am indebted to both authors for having given me pre-publication information about the latest names, so that these might be included in the third edition, which followed shortly after publication of their works. Details of these monographs and other scientific works are listed at the end of this book for those enthusiasts who wish to do further reading.

The forerunner to the taxonomic revision of the genus *Protea* by Dr. Rourke, published in 1980, has resulted in several name changes, which have been incorporated into the sixth edition.

In this new edition, *Proteas for Pleasure* retains its aim, which is to simplify the complicated Protea family for the amateur and highlight its most attractive members, both for the gardener and the florist.

SIMA ELIOVSON

PROTEA—FLOWER OF A NATION

The Protea has become South Africa's floral emblem by common usage. At one time, the White Arum Lily (*Zantedeschia aethiopica*) was proclaimed as the national flower, but the striking Protea has so fired the public imagination that it has been regarded without question as the nation's most typical flower. This natural selection has been upheld by officialdom and the Protea is now recognised as the floral emblem of South Africa.

The Suikerbos, as it is termed affectionately, is used as an emblem for many things. *P. caffra*, which is a very commonly seen Protea, was used on the "tickey" and sixpence, both small silver coins which are now equivalent to $2\frac{1}{2}$ and 5 cents respectively.

The term "Suikerbos" or "Sugarbush" is used to denote all Proteas, but the true Sugarbush that originately acquired the name was *Protea repens* (formerly known as *P. mellifera*). This flower has more abundant nectar than any other species and was used as a form of sweetening by the early colonists, as well as for medicinal purposes. They distilled a potent syrup called "Suikerbos stroop" or "Bossiestroop" from it. This was one of the first Proteas to have been seen and admired by the early visitors to the Cape.

Although all Proteas are regarded equally as South Africa's national flowers, *P.repens* is the most important species traditionally. As such, it is the most suitable choice if only one species were to be selected to represent the genus. To quote R. Gerard in *Flags over South Africa* (1952), "The Protea (*P. mellifera* Thunb.)—the Sugarbush—was the symbol of the Cape of Good Hope and is now accepted as the national floral emblem of the Union of South Africa." *P. repens* was used during the Golden Jubilee celebrations of Kirstenbosch in 1963 as the emblem of "South Africa's Floral Year".

During 1976, the government decided that the King Protea, *Protea cynaroides*, was to be chosen as the official floral emblem of the Republic of South Africa.

INTRODUCTION

Whatever you may feel about Proteas, it is fairly certain that you cannot be indifferent to them. You may like them for their own beauty or you may dislike them, but you cannot be completely impervious to them. An Englishman to whom I spoke about Proteas, said, "What! Proteas! Great big vulgar things! How can you possibly like them?" He was, of course, thinking of the King Protea (*P. cynaroides*) which is, to my way of thinking, breath-taking in its magnificence and perfection. Had I been able to show him some of the tiny Proteas, like *Protea odorata*, a sweet-scented miniature, scarcely larger than a cornflower, or the dainty Mountain Rose (*Protea nana*) with its nodding rosy head about the size of a plum, he might perhaps have been surprised and pleased to meet these little Proteas and to learn that there are many sorts of these fascinating flowers. For Proteas are a whole vast tribe in themselves and you may choose your favourites from a multitude of types.

A strange phenomenon of taste is the way in which it can be developed. I am deeply conscious of how my own love of Proteas has grown, as has that of many people whose interests differ greatly from one another. Familiarity, of course, leads to appreciation. If you hear the familiar strains of a piece of music, you respond to the intimate knowledge of every phrase and note. You love what you know and have learned to distinguish for yourself. Naturally, the music must have had an attraction in the first place, but you can learn to love almost any good music provided that you hear it often enough so that you can recognise it as an old friend.

Proteas, too, become more and more beautiful and fascinating with increased awareness of them. They are seldom flowers that you fall in love with at first sight as you might do with a rose or a daffodil. But then there must always have been roses or daffodils somewhere in the remembered past, whereas the impact of Proteas could be quite a new experience. For Proteas are seldom seen outside their country of origin, as they have not yet become common in gardening or commerce, owing to their exclusiveness, both in culture and availability.

Proteas come only from Africa, being particularly concentrated in the southern tip of the continent, and unless you have seen them picked freshly from the bush, you will never be aware of their subtle appeal. Because Proteas last for a long time and dry naturally in the vase, this does not mean that they are at their best at any time. The best time to fall in love with a Protea is when it has reached its unblemished peak of perfection, when it has probably been freshly picked and arranged in the vase.

Unfortunately, familiarity also breeds contempt, and in the past South Africans have been known to look down on the Proteas which are so commonly seen in the Cape. That was in the days when one saw mainly the spiky *Protea repens*, commonly called the Sugarbush or Suikerbos, which was the best known example of the tribe. This species has a long shiny bud, sticky with nectar, opening into a cup that has a hard appearance once it has lost the exquisite blush of colour that is present when it is fresh. Thinking only of this Protea, it might be said that Proteas are stiff and hard. You could not make that statement, however, knowing the Giant Woolly-beard Protea (*Protea magnifica*) which resembles a large glamorous powderpuff, whether it be the lemon yellow colour variety, the boudoir pink or watermelon red. The satiny pink bracts of *Protea speciosa*, tipped with rich brown fur, or the striking *Protea neriifolia*, with its black furry rims and its dark hairy cone at the centre, are both soft and graceful in their appearance.

The truth is that there are Proteas to suit all tastes and that you can enjoy the differences between them in addition. The pleasure of a bowl of mixed Proteas is heightened by the contrast between them. You might not like the apple green *Protea coronata* by itself or the brownish pink *Protea susannae*, but they help to weave a subtle tapestry of colour tones in a vase where the more dramatic and brilliantly coloured Proteas are the stars.

Though we have discussed only Proteas, we must remember that Proteas themselves are the chief members of a large family which is known as the Protea family (*Proteaceae*). There are many outstanding beauties in the other groups in this family. Notable amongst them are the magnificent Pincushion Flowers (*Leucospermum*) which have wonderfully constructed heads of flowers that resemble a bristling pincushion. Whether the "pins" have a gleaming or a velvety texture, they glow with colours ranging through oranges and salmons to rosy pinks and are irresistible both in the garden and the vase.

One of the smaller groups in the Protea family is *Serruria*. Its most famous member, the Blushing Bride (*S. florida*) is, possibly, the one species which would appeal immediately to anyone as a flower, without having been initiated into the intricacies of Proteas. Its dainty nodding blooms are best seen in a specimen vase where one can admire, at close quarters, the waxy white pointed bracts, delicately flushed with pink or palest green, wide-spread to reveal the hairy, pink-tipped centre.

LEARNING TO RECOGNISE PROTEAS

Studying Proteas as a hobby can bring pleasure to many—to the gardener, to the amateur arranger of flowers or florist, to the nurseryman or grower for the cut flower trade, as well as to the person who values the floral heritage that belongs to Southern Africa.

Studying Proteas also presents a difficult problem to the layman who wants to learn how to distinguish between them. It is possible, however, to learn the differences between them without being a botanist or knowing more than the simplest botanical term. Most of the amateur botanists, including the author, who have learned about Proteas and are still continuing to learn about them, have simply found out how to recognise them by familiarity; by being told what it is that makes the botanist distinguish between them, and by having developed their observation powers to note small details.

It may come as a surprise to know that the botanist does not recognise colour as a clue in the differentiation between species of Proteas, partly because the colours vary so much in each species. It is the botanical features like the shape of the bracts and the presence of hairs that count, as well as the whole appearance of the flower head.

It is necessary to know that a Protea is not a flower with petals like an ordinary flower, but a collection of small flowers that make up a composite flower-head, like a daisy. The individual flowers are dull and would not be looked at twice in the normal way, but they are grouped in a mass to form a hairy cone or bristly brush, being surrounded by petal-like bracts which are variously coloured or befurred. These bracts, together with the mass of individual "flowers", make up the "flowerhead", which we refer to familiarly as a flower.

Work goes on continuously at a scientific level in the study of any group of plants, and the complicated Protea group is no exception. The amateurs depend on the knowledge assembled by the scientific botanist, who studies the intricacies of the correct names, the distribution or area over which the plants are found in the wild, their historical references and many other factors that add to the sum of knowledge. Proteas present problems even to the scientists. The lesser-known ones are sometimes so rare that the expert has still to discover details about them. The Leucadendrons, for example, are still imperfectly understood and there are many fascinating discoveries still to be made before the jigsaw puzzle can be completed.

If one were to wait until the last word had been said on the subject, when all the details were finalised and published, then the amateur might have to wait for many long years for a layman's book to be written. The layman is not inhibited by the exact demands of science and the protocol attached to the publishing of new names as a result of research, so that he can go ahead and assemble what is known up to the time of writing. Thus it is that fools rush in where angels fear to tread and the author has attempted a task from which the expert might well flinch. It is, however, with a sincere love of the Protea family that I have endeavoured to present the beauty of the Protea to a public that might benefit from seeing the pictures and learning to recognise the flowers, as well as from hearing more about growing these plants and of their value as cut flowers. The scientific facts have been sought out because it is necessary to have them as a basis for recognition and understanding. After that one must grow the plants and pick the flowers until they become so familiar that they are part of one's fibre and one recognises them instantly without thinking.

This book then is offered in humility to those who love Proteas and to those who want to learn about Proteas, in the hope that the scientific person will enjoy the assembled photographs and the layman will enjoy both the pictures and the details about the individual species. It is hoped that both will make use of the horticultural information picked up after years of mixing with Protea growers and the friendly folk who make up the gardeners of the world. My aim is to arouse an interest in Proteas, which are not only our national flowers, but our heritage.

PART 1

CHAPTER I

PROTEAS AND THEIR NAMES

Amongst the legends of ancient Greece is a tale about Menelaus, King of Sparta, who tried to help Telemachus find his father Odysseus, who had been delayed on his way back from Troy. He told the young man that he had managed to capture the Greek god Proteus, also known as the old man of the sea, who would foretell the future if he were caught. Proteus changed into every imaginable shape in order to avoid being caught, but Menelaus struggled and managed to retain his grip while Proteus took on one shape after another. Finally he overcame him and forced him to reveal the destinies of the returning heroes. Proteus revealed that Odysseus was the captive of the nymph Calypso, who kept him prisoner on a lonely island.

The Proteus of many shapes was the mythological figure that sprang to the mind of Carl Linnaeus, the great Swedish naturalist, when he named the Protea. His classification system was adopted throughout the world and marks a milestone in the naming of plants. Up to this time plants were given long Latin descriptions and he devised the method of referring to them by two names only, that of the genus and the species.

Linnaeus' choice of the name *Protea*, in 1735, was a stroke of genius, for it conjured up immediately to the well-read person what a variety there must be in this genus or group of plants. Even in his day there were scores of species of *Protea*, while many more have been discovered since that time, even in very recent years. Later, in 1789, Professor Jussieu of Paris gave the name *Proteaceae* to the whole family, naming it after the most outstanding member, the genus *Protea*.

The seemingly annoying habit of plants having their botanical names changed is the result of a decision by later botanists to recognise only the first published name as from the time of Linnaeus. Before his time, different botanists often gave different names to the same plant and this led to confusion. Constant research and revision by modern botanists often reveals that plants which were named after the time of Linnaeus were renamed by botanists who had no knowledge of earlier discoveries. The international rule was made that, in all cases, the earliest name since the time of Linnaeus must be used, or "upheld", to use the terminology of botanists. It matters not that the later name might be a better one, because to deviate from the rules would lead to all sorts of exceptions.

As a result of the recent research by Professor H. B. Rycroft in revising the genus *Protea*, it was necessary to change the charming name of *P. mellifera*, meaning the honey-bearing Protea, to that of *P. repens*, meaning the creeping Protea. This is an unfortunate misnomer, due to the fact that Linnaeus probably named it from an illustration, never having seen the lovely large bush in nature. Thunberg renamed it *P. mellifera* some years later, but rules must be obeyed and we must now refer to our True Sugarbush as *P. repens*, realising that it is sometimes uncomfortable to understand a little Latin.

SCIENTIFIC AND POPULAR NAMES

Most people who are interested in plants and animals know that scientific and botanical names are given to them so that they may be known throughout the world by these names. The knowledge of this international language helps one to communicate with people all over the world who have the same interests. There are many types of the same plants and animals throughout the world and to know the name of the genus or group helps one to recognise their relationship. To refer to them by local names would be meaningless, while common names which are charming and easy are no guarantee that one is referring to the correct plant. Local people often give the same name to many plants. There are hundreds of daisies, for example, but you would not know which you wanted if you did not know its botanical name. It would, of course, be too pompous and pedantic to use only the botanical names for plants, but one should try and learn both so that the correct plant can be ascertained if necessary.

When people have grown up for centuries in a land with few wild flowers, it is easy to spread the use of their common or popular names. Well-known garden flowers are known by their common names rather than by their botanical names, such as Poppy for *Papaver*, Snapdragon for *Antirrhinum*, Lavender for *Lavandula* and many others. South Africa is too young, its population too spread out and its wild flowers too plentiful for all of them to have become familiar enough to have acquired common names. Plants at the Cape, where people first made their homes 300 years ago, have many common names such as "Kalkoentjie" (meaning "Little Turkey") for *Gladiolus alatus* and "Gousblom" (meaning "Gold Flower') for *Arctotis fastuosa*—a type of daisy with orange flowers. This name has also been given to other types of daisies like *Dimorphotheca,* as well as to the cultivated *Calendula*. It can be seen, therefore, that a common name is not dependable.

Nevertheless, without popular names, many plants would not become popular in horticulture, for most people find it difficult to master the scientific names. It is not only important to know the established common names, but to invent everyday names that will help to popularize beautiful plants. When plants were described for the first time in that famous collection of botanical art, Curtis's Botanical Magazine, dating from 1790 to the present day, the botanists translated the descriptive Latin names into English. *Protea mellifera*, for example, was called the Honey-bearing Protea. Later, the colonists gave it the name of Suikerbos (Sugarbush), when it had become so familiar that a popular name grew up as a result of someone's imagination.

Hundreds of South African wild plants have no popular names as yet, and it is a pity that they should remain anonymous in the everyday world, waiting for a name to grow up through usage. With this in mind, the author gave some wild flowers common descriptive names several years ago and has been gratified to hear them referred to colloquially by these names. "Gold-Tips" for *Leucadendron xanthoconus* and "Rose Cockade" for *L. tinctum* have found their way into other books and are now established by usage, so that this step was justified. By calling *Leucadendron platyspermum* the "Knobkerrie Bush", one feels that it must become better known to the public. One could carry on, with restraint, hoping always that the scientific names will be used in preference by those who take the trouble to learn them.

One must also accept the fact that business men influence the public and give horticultural names to plants in order to popularize them. Wild flower lovers who genuinely wish to popularize indigenous flowers will not decry their efforts. The common names of King Protea and Queen Protea have long existed for *P. cynaroides* and *P. magnifica*, respectively. The names "Prince" for *P. speciosa* and "Princess" for *P. grandiceps* have been mooted by an exporter of cut flowers and it remains to be seen whether these will be accepted by the layman. Purists may object, but if a rose can be popularized as "Peace", why not a Protea? The object should be to make Proteas known to the world and to encourage gardeners to grow them, so that any pleasant method of doing this should be encouraged.

Rudolph Marloth, South Africa's first trained botanist and a great man, wrote a book called *The Common Names of Plants* (published in 1917 and now out of print), in which he stated that "common names . . . certainly help to disseminate knowledge and love of plants among the people, and that alone would justify their preservation."

THE PROTEA FAMILY

The ordinary gardener finds it difficult to understand the differences between Proteas and the Protea family. By having the name *Protea* for the single genus or group of plants and the name *Proteaceae* for the whole family, things are so simple that it confuses those who do not know the difference between a genus and a family. The fact that we often refer to all members of the Protea family as Proteas adds further to the confusion.

The word *family* refers to a large collection of plants with similar characteristics, comparable amongst humans to a race of people. The *genus* refers to a particular group in the family and the generic names given to these groups are equivalent to the surnames given to families of human beings. There are many species in each genus and the specific names given these are comparable to the Christian names given to individuals. Furthermore, there may be varieties of a species. This may be a colour variety or a small characteristic that sets it apart, but it is not sufficiently important to warrant it being a separate species. It is, however, a distinct sub-division of a species growing in the wild.

The way to refer to a plant by its full scientific name is to name the genus first, followed by that of the species, e.g. *Protea nana*. There are very few varietal names amongst Proteas, although varieties of the same species often exist. Colour forms of Proteas do not have scientific names as a rule. One is permitted to use the initial for the generic name *Protea*, provided that it has been mentioned in full at the commencement of writing.

The Protea family (*Proteaceae*) refers to the large race of which *Protea* is a single group or genus. Other members of the family are the genera known as *Leucospermum, Leucadendron, Mimetes*, etc., each with their own collection of species. Altogether, there are 61 genera and about 1,400 species in the Protea family.

PLANT GEOGRAPHY OF THE PROTEA FAMILY

The Protea family occurs chiefly in the Southern Hemisphere, especially in Australia and South Africa, with a few in Tropical Africa, South America, Tropical America, Malaya, New Zealand and the Pacific Islands. There is a theory that Africa and Australia were once linked together and broke away in prehistoric times. The fact that the Protea family is so richly represented in both continents shows a relationship in the flora of the Cape Region and that of Australia and is one of the reasons in favour of this theory.

Plant geography is an interesting subject and when we discover where plants grow wild it often helps in understanding how to grow them.

There are 14 genera of the Protea family in South Africa, which are all exclusive to Africa and none grows wild anywhere else in the world. The genus *Hakea*, of which 5 species have naturalised themselves in South Africa and grow wild in parts of the Cape, are Australian importations. These, as well as some Acacias, *Pinus pinaster* and Australian Myrtle, are termed "alien vegetation" by wild flower protectors, for they strangle the natural vegetation by virtue of their prolific germination and vigorous growth and are eradicated from all wild flower reserves wherever possible.

Most species in the Protea family are concentrated in the S.W. Cape, where they help to cover the mountainsides with attractive, evergreen vegetation known as "fynbos". This is the short, bushy, ericoid vegetation of the winter rainfall area of the S.W. Cape, as distinct from grassland. It is equivalent to the "macchia" or bushy vegetation of the Mediterranean in Italy, the South of France and Spain.

The mass of Proteas grow mainly in the mountains which curve around the coastline of S. Africa, spreading in a rough crescent from the Clanwilliam district to the area around Port Elizabeth, with its thickest portion in the Tulbagh area, the Peninsula, Caledon and Knysna. A smaller number of Proteas grow in the eastern mountain ranges, extending northwards

MAPS SHOWING PLACE NAMES
The place names are those mentioned in the book to indicate the areas in which Proteas are found growing wild.

into Natal and the Transvaal, with a few occuring in Tropical Africa. (See Maps.)

MEMBERS OF THE PROTEA FAMILY THROUGHOUT THE WORLD

The following is a list of the chief members of the Protea family, separating those which are indigenous to Africa from those which occur in the rest of the world.

MEMBERS OF THE PROTEA FAMILY IN AFRICA

Aulax (3 species in the S.W. Cape)
Brabejum (1 species—S.W. Cape)
Diastella (5 species—S.W. Cape)
Faurea (5 species—Cape, Natal, Transvaal, 1 sp. in Malagasy)
Leucadendron (81 species—mainly S.W. Cape)
Leucospermum (About 47 species—mainly S.W. Cape. A few in the Transvaal, Natal and Tropical Africa)
Mimetes (16 species—S.W. Cape)
Orothamnus (1 species—S.W. Cape)
Paranomus (17 species—S.W. Cape)
Protea (about 114 species—mainly S.W. Cape. Also Natal, Transvaal, Lesotho, Swaziland and Tropical Africa)
Serruria (50 species—S.W. Cape)
Sorocephalus (11 species—Cape)
Spatalla (20 species—mainly S.W. Cape) (Includes former genus *Spatallopsis*)

The last two genera have small flowers and are usually ignored as they are not well known.

SOME MEMBERS OF THE PROTEA FAMILY IN OTHER PARTS OF THE WORLD

Adenanthos (W. Australia)
Agastachys (Australia)
Banksia (Australia)
Conospermum (W. Australia)
Dryandra (Australia)
Embothrium (Chile)
Gevuina (Chile)
Grevillea (Australia, New Caledonia)
Knightia (New Zealand, New Caledonia)
Lambertia (Australia)
Lomatia (Australia, Chile)
Macadamia (Australia)
Persoonia (Australia, New Zealand)
Petrophila (Australia)
Roupala (Tropical America)

Hakea (Australia)
Hicksbeachia (Australia)
Isopogon (Australia)
Stenocarpus (Australia, New Caledonia)
Telopea (Australia)
Xylomelum (Australia)

POPULAR QUESTIONS ABOUT PROTEAS

People frequently ask the following questions:
1. How many Proteas are there?
2. How many members are there in the Protea family?
3. Do Proteas grow wild in other parts of the world besides Africa?
4. Do Proteas grow only in the Cape?

ANSWERS

1. There are about 114 species of Protea up to the time of writing. New species often appear during the process of sorting and reclassifying, apart from actual discovery in inaccessible parts of the mountains in the Cape, as well as in the summer rainfall area.
2. There are 61 genera or members in the Protea family, occurring chiefly in the Southern Hemisphere and mainly in South Africa and Australia. There are about 1,400 species in the Protea family.
3. Proteas do not grow wild anywhere but in Africa, occurring chiefly in southern Africa and especially near the tip of the continent in the S.W. Cape. Members of the Protea family, however, grow wild in other parts, especially in Australia.
4. Proteas do not grow only in the Cape, although there was once a belief that they could not be cultivated away from the Cape. Proteas have been cultivated successfully in all parts of South Africa. They may be seen thriving in all provinces, from sub-tropical, coastal Natal to the temperate inland gardens of the highveld, with temperatures dropping occasionally to zero in winter.

The majority of Proteas grow wild at the Cape, especially the most attractive kinds. Proteas also grow wild in the other provinces, particularly in the mountainous parts of Natal, the Transvaal, Zimbabwe and Tropical Africa.

CHAPTER II

ECHOES OF THE PAST

It is fascinating to look back into the past and to learn how members of the Protea family were studied, appreciated and grown by early plant collectors, ever since the year 1605, when the first Protea was described and illustrated in a work by the French botanist, de l'Cluse, known as Clusius.

This was *Protea neriifolia*, although at the time it was not known as a Protea and was, indeed, referred to as a graceful kind of thistle. It was also mistakenly thought to have come from Madagascar, but this was later disproved by researchers. It is interesting to note that this was not only the first Protea to be described, but also the first South African native plant to be recorded in literature. With this in mind, it is more than fitting that the Protea should be recognised as South Africa's national flower.

The age of discovery of the 16th and 17th centuries stimulated exploration of the newly discovered countries and led scientists, traders and missionaries to bring back treasures to their patrons, as well as spread civilisation. The wealthy owners of gardens in Europe, including royalty, sought to build up collections of rare plants from foreign lands, and the rich flora of the Cape was especially valued. Plant collectors were sent out from Europe, starting in 1772 with Francis Masson, who was the first official collector sent from Kew. He travelled to many countries, but paid two visits to the Cape, of two and ten years respectively, making major collections of wild plants that were sent back to Kew for cultivation and classification. Proteas formed an important group in his collection.

The explorers and plant collectors who came to the Cape are famous throughout the botanical world and their names are reflected in the names given to many of our native plants. Great botanists and botanical travellers like Robert Brown, Carl Thunberg, James Niven, Oldenland, Koenig, Sparrman, Hermann, Paterson and Masson have left their names and influence on our flora. Botanical collectors of the 19th century like Ecklon, Carl Zeyher, J.F. Drège and Carl W. L. Pappe contributed to our present-day knowledge and their names have a familiar ring to those who know something of the South

African flora. The noted naturalist, William J. Burchell, not only collected plants, but made drawings and notes during his travels for four years through the country by ox-wagon.

The history of plant collecting is full of romance and adventure and the men who undertook it had a love of wild nature that made them endure physical hardship in the excitement of contributing to the world's knowledge of flowers. Many of the early travellers were cultured men of science, doctors, botanists and authors, all with a love of adventure and a love of plants. The traveller of today, who is interested in wild flowers, can see them almost from the roadside at the right time of the year and can reach almost any place in the country by car.

Botanical authors were inspired to write up the South African flora. Robert Brown, the British botanist, published important studies *On the Proteaceae of Jussieu* in 1810, after Professor Jussieu of the University of Paris, an authority on classifying plants, had given the name *Proteaceae* to the whole family in 1789. Linnaeus, who reduced long Latin descriptions of plants to two words, that of the genus and the species, had previously given the name *Protea* to the most important member of the family in his *Systema Naturae* in 1735. Thunberg, his pupil, listed the authors who had written on South African plants, namely, Plukenet, Boerhaave, Hermann, Breyne, Commelin, Burmann, Linnaeus, Berg and Jacquin.

Carl Friedrich Meisner (1800-1874) brought all the knowledge of Proteas up to date, but it was not until 1910 that Dr. E.P. Phillips was sent to Kew to study all the collected data from all over the world, which he finally published with Dr. O. Stapf in *Flora Capensis* (V. Sect I., 1912). A new flora of Southern Africa is now being compiled, which will take about 25 years to complete. Professor H. B. Rycroft, present director of the National Botanic Gardens at Kirstenbosch, is working on the genus *Protea*. Since this book was first published, three important genera have been revised, namely, *Paranomus,* by Dr. M. Levyns, *Leucospermum*, by Dr. J. P. Rourke and *Leucadendron*, by Dr. Ion Williams. Botanic research in Protea cultivation is being continued by Mrs. M. Vogts of Betty's Bay. It may be seen, therefore, that the work of botanic research goes on continuously and that the last word has not yet been said on the Protea family in South Africa, apart from the remaining members in other parts of the world.

The story of South African botanical literature is long and

interesting. An impressive summary of it is given in John Hutchinson's book, *A Botanist in Southern Africa*. It is interesting to note that nearly every author made some mention of the fascinating Protea family.

Many specimens of Proteas were taken to Europe by the turn of the 18th century. Although very few were described in those early days, many were illustrated. The Wild Almond, *Brabejum*, was one of the first to be illustrated in 1648 in a work of Jacob Breyne. Herman Boerhaave illustrated 23 members of the Protea family in 1720, using specimens and drawings sent to him by Hartog, the head of the Dutch East India Company's garden at the time. The pages of Curtis's *Botanical Magazine*, started in England in 1790, were filled with magnificent paintings of flowers, many of them South African, including many Proteas. Hendrik Claudius was an artist employed by the Dutch East India Company to paint wild plants. H. Andrews in his *Botanists' Repository* (1799-1811) painted beautiful coloured pictures of new plants in British gardens, including many species of Proteas. R. Marloth's *Flora of South Africa* is filled with magnificent paintings of South African wild flowers. These treasures may be seen in large libraries throughout the world.

Many drawings of Cape plants were made by laymen, who were fascinated by their beauty. Mrs. A. E. Roupell, who stayed at the Cape between 1842 and 1845, painted numerous wild flowers of the country chiefly to please her husband. Several plates were published in 1849 by the director of Kew under the modest title of *Cape Flowers by A Lady*. Among these eight beautiful illustrations were at least three members of the Protea family. Some of these have been republished recently.

It is not only the botanist who has contributed to the knowledge of Proteas. It was the gardener who made them live in the early days when Proteas were grown in Europe. Skilled nurserymen of Europe, like Joseph Knight, who was the gardener to George Hibbert of Clapham and who was responsible for making Hibbert's the largest living collection of Proteas in the world, published a book on their cultivation in 1809. He grew them in drums, where they flowered and set seed, bringing them into greenhouses for shelter during winter. Many individual plants were also grown in European centres like Vienna, Berlin and Amsterdam.

The first members of the Protea family were planted in South Africa in 1660, when Jan van Riebeeck enclosed the new settlement with a hedge of Wild Almond, *Brabejum*, part of which still remain today above Kirstenbosch and are preserved as a national monument. Silver Trees were used as windbreaks in the Dutch East India Company's garden. The gardeners in charge of the Company's garden contributed much to the study of Proteas and other South African flowers. Hartog, the head of the Company's garden at the turn of the 18th century, sent specimens and drawings to Herman Boerhaave for his work published in 1720. A work on the flora of the Cape, describing 17 members of the Protea family, was published in 1767 by Bergius of Sweden, who received his specimens and information from Auge, then head of the Company's garden at the Cape.

Proteas were not much cultivated in South Africa during the succeeding years, partly because the colonists looked upon local flowers as weeds and partly because the flowers could be picked so freely in the fields. When plentiful flowers became rare through picking and wild flower protection was recognised in the 20th century, people began to cherish their floral heritage and wish to preserve their flowers by growing them in their gardens.

Knowledgeable South African nurserymen of the past twenty years have made a tremendous contribution in popularising Proteas, making the plants available to the gardener and the flowers to the florist. There are many specialist Protea nurseries in the Cape as well as in Natal and the Transvaal, but it is almost true to say that most good shrub nurseries stock Protea plants and the present-day gardener is fortunate in being able to purchase well-grown plants from his local nursery.

The example set by wild flower gardens all over the country is an inspiration to those who seek the beauty of Protea plants without having to explore the mountains of the interior like the plant hunters of old. The National Botanic Gardens at Kirstenbosch, Cape, attract thousands of visitors each spring to see the Proteas as well as other wild flowers. Caledon Wild Flower Garden in the Cape is a mecca for wild flower lovers. The "Karroo Garden" at Worcester and the "Harold Porter Botanic Reserve" at Betty's Bay both belong to Kirstenbosch and are at their best in spring. "Settler's Park" in Port Elizabeth, "The Wilds" in Johannesburg, Pretoria National Botanic Garden —all these wild flower gardens give untold pleasure to the visitor and visible proof of the ease with which Proteas can be grown in cultivation.

The everyday gardener who grows Proteas can do more to stimulate others than any public garden, however, for there is no more eloquent appeal to plant a Protea than to see one flourishing in one's neighbour's garden.

The ordinary nature-lover has contributed a great deal to the knowledge of Proteas to this day. Those who roam the mountains and see the flowers in their natural surroundings, bringing discoveries of new forms and even new species, are of the greatest help to the writer and the botanists who labour at their desks in their work of research. Enthusiasts like T. P. Stokoe, whose name is immortalised with *Protea stokoei*, climbed the mountains till his death in 1959 at the age of 91, and discovered many new plants which he brought to the botanists of Cape Town. New discoveries have been made even more recently by mountaineers and nature-lovers, so that this great work of discovery continues to this day, while the present merges into the past and the grains of information add up to a bold pattern in the scheme of nature and knowledge.

CHAPTER III

PROTEAS RARE AND COMMON

Proteas which are common in nature are not always commonly known to the average person, while plants which are common in commerce may become, through their very popularity, rare in nature.

In Mrs. Roupell's book on the Cape flora, published in 1849, *Protea repens*, then known as *P. mellifera*, was pictured as very prolific, and described as follows: "Whoever has visited South Africa . . . includes the beautiful shrub . . . among the most vivid of his recollections of the Cape flora, for this is one of the first of the native shrubs that catches his eye on landing; and wherever afterwards he may wander through the Colony, it accompanies his steps." It was certainly commonly known when it was given its popular name of Suikerbos or Sugarbush. A folk song was written about the Suikerbos, which is famous to the present day. The word Suikerbos is now given to all Proteas, although it was meant originally only for the nectar-sticky *P. repens*, which was used for sweetening by the early colonists.

Today, however, *P. repens* is not so very common in nature, and the areas in which it grows have diminished through the enemies of wild flowers—the spread of farming, fires, road-building, flower picking and the general penetration of civilisation. It is still well-known at the florist, as it is one of the most easily grown commercial species, but it might well be that *P. repens* may become one of the rarer Proteas in nature as time progresses.

On the other hand, plants that have been rare in nature and, in fact, completely lost to science, have been brought back into the fold by cultivation. The Blushing Bride (*Serruria florida*) is the best example of the rescue of a member of the Protea family, for it was rediscovered in the Fransch Hoek mountains of the S.W. Cape after 90 years of seeming extinction. The plants were guarded by Kirstenbosch until the seeds ripened. The seedlings which were then cultivated by Kirstenbosch have been the parents of the plants which are now so feely obtainable in commerce. *Protea aristata*, which was so rare in nature that only a handful of plants were known, has recently been made available in the trade, and its lovely flower may become as well-known as the Blushing Bride. It is even possible that more plants may be re-introduced into their native haunts in the future by conservation-minded citizens in the areas where they are found.

Beautiful Proteas like *P. magnifica* (Giant Woolly-beard), *P. cynaroides* (King Protea) and *P. speciosa* (Brown-bearded Protea) are seen much more frequently as cut flowers than they were a few years ago, and popular interest has grown so much that very unusual kinds are now finding their way into our vases. The ubiquitous *P. compacta*, that is sold on every street corner stand, is extremely well-known, and it is difficult to believe that this may become rare through indiscriminate picking or lack of conservation.

Cultivation does not always provide the answer to saving really rare Proteas, however. The strange and interesting Marsh Rose (*Orothamnus zeyheri*) is one of the few members of the Protea family that is threatened with extinction in our lifetime. It grows in small, isolated patches in the southern Hottentots Holland Mountains at the Cape, in places where few people wander. It has a natural tendency to die out and, should the flower be picked with a long stem, the whole slender plant will die back altogether, leaving a dry dead stalk as evidence of its existence. Many efforts to cultivate it have ended in failure. It has been grown from seed and even brought to the flowering stage, yet it seems to resent capture and the enthusiast is faced with the fact that some secret is necessary to keep the Marsh Rose with us, even in nature.

Mountain climbers and nature-lovers ramble into the higher slopes of the mountains and discover treasures to this day. Proteas that are still unnamed and must wait for scientific tabulation before they are recognised as individual species have been photographed in high altitudes where snow falls, while unusual varieties of known species sometimes occur unexpectedly in outlying districts.

With Proteas, therefore, we are still living in an age of discovery and it is exciting, as one delves further into the complex subject, to realise that the last word has not yet been said on the Protea. Let us not allow our familiarity with some well-known Proteas to dull our senses or lead to boredom, for it is important to nurture the whole group so that we may protect them in nature and enjoy them in our gardens.

CHAPTER IV

ON GROWING PROTEAS

When plants such as Proteas appear sometimes to be unpredictably difficult to grow, speculation is always rife among gardeners about the ideal requirements that they should be given in the garden. Secrets of success are exchanged and failures are explained away "with every imaginable reason from disease to planting at the wrong phase of the moon," to quote a puzzled Protea enthusiast in California.

The serious grower will obviously try to analyse the natural conditions of growth; to study the places where Proteas are found growing wild in order to seek the clues that might lead to success in cultivating them. These studies are helpful, but the gardener should not be discouraged by the fact that he might not have these conditions in his garden, for it often happens that Proteas perform better when cared for in gardens than they do in the wild, even when they are grown in conditions that are in complete contrast to those of their natural surroundings.

I never cease to be amazed at the way Proteas thrive and flower in inhospitable conditions, in poor sub-soil, without care and often with the minimum of water. Such results give rise to the claim that Proteas do not require water or good soil. The fact that they do even better in good soil and with regular watering, however, points to the fact that one need not mistreat them in order to grow them well. Resistance to drought and tolerance of poor soil are simply encouragements to those gardeners who do not have naturally good soil or ample facilities for watering. The dwarf species with underground stems are particularly able to withstand drought.

The circumstances in which Proteas grow in nature vary so much that one has a great deal of leeway, yet one can draw some conclusions that point the way to achievement.

THE SOIL

If I were to state the most essential factor in the cultivation of Proteas it would be that they should have well-drained soil. The fact that they grow wild chiefly on mountainsides in nature is indicative of their preference for water run-off.

Proteas are not found in marshy ground and quite often grow in sandy places. A few Leucadendrons, like *L. uliginosum* and *L. salicifolium,* sometimes grow on the banks of rivers, but these are exceptions rather than the rule. Proteas grow in many different kinds of soil in nature, but these are generally well-drained.

One of the theories put forward in favour of well-drained soil is based on recent scientific research by botanists, notably Mrs. M. Vogts, which revealed that Proteas develop an autogenous toxin in the husk of their seeds and in their root system. This natural poison inhibits germination and often kills a well-grown healthy plant that dies suddenly for no apparent reason. It is thought that this toxin can be leached or washed out of the soil by a continuous run-off of water and that good drainage is therefore essential so that the toxins can be washed away during watering.

The ideal type of soil should have a gravelly character so that water drains off well. To test your own soil, dig a small hole and pour in a bucket of water. If the water sinks down fairly quickly, the soil is ideal. If the water remains in the hole for a long time, making a muddy puddle, then it is not likely to be a good spot to grow Proteas. Shale or clay soils are not recommended, and one seldom manages to grow Proteas in such soils. Proteas can be grown in shale soil, however, if the gardener chooses a sloping bank or the top of a rockery and makes an effort to correct drainage.

HOW TO CORRECT POOR DRAINAGE

Badly drained soil may be natural or man-made. When constructing homes on hillsides, soil may be rammed down for stability and this will cause compacted soil. Natural poorly drained soil is composed of heavy clay or shale or soil with an underlay of rock.

Sometimes one can dig through to a porous layer of soil, in which case one's problem is solved. Machinery can be used if a shovel is inadequate. At any rate, the soil should not be worked if it is too wet or too dry. If it is too dry, soak it two or three days before attempting to dig the area. The planting hole for Proteas should be at least 3 or 4 feet wide and 3 feet deep. If one cannot dig the hole to connect down to a porous layer, one can dig a foot-wide "chimney" from the bottom of the planting hole to a porous layer. This chimney should be filled with fine soil, or peat-moss or either of these materials mixed with the existing soil. Fine sand is said to attract water more than large rocks or gravel and the water will sink down to the sand by capillary action, whereas the hole may become water-logged before gravity will force the excess moisture down towards a rocky base.

If there is no porous subsoil, one can remedy poor drainage in other ways. Make the hole even wider and deeper and fill it with porous material. The bottom of the hole should be filled with rough twigs and coarse sand and the top half filled with light sandy soil mixed with well-rotted compost. It is important to give the soil a fibrous quality and one could add peat-moss or even sawdust, treated with nitrogen, to lighten the texture. It is especially important not to overwater the plant in such a situation.

One can construct a raised bed in the form of a rockery or edge it with a retaining wall with "weep" holes at the base, if it fits in harmoniously with the general lay-out of the garden.

Trenching is one of the best methods of draining away water over a large area. Dig a deep trench and fill the base with stones, covering these with soil. The Proteas can be planted normally above the trench. This is an elaborate procedure which is used only in places with water-logged soil.

Underground drainage systems are expensive and sometimes there is nowhere for the water to go, so that these are not always practical. Drain tiles can be placed underground when there is a slope in the soil.

In places where it is very difficult to correct drainage, one could grow the smaller Proteas in large pots or tubs and use them decoratively on a patio. (See Chapter on Landscape Design).

HOW TO PREPARE THE PLANTING HOLE IN NORMAL SOIL

It is not necessary to add very much to a naturally well-drained or gravelly soil, but one should always prepare a hole at least two feet wide and two feet deep before planting even the smallest seedling. Whenever possible, mix plenty of rough vegetable material, compost or humus into the lower half of the hole. If one does not have sufficient compost, break up the soil to loosen it, in any case, and try to mix in some twigs and rough plant material such as autumn leaves, even if they are not decomposed, as they will moulder by the time the Protea roots penetrate to their depth, and can only benefit the plant. Reserve whatever good leafmould or well-rotted compost that is available for the top of the hole, mixing as much as possible into the top four or six inches. This soft-textured mixture will assist the young roots to spread easily, while the organic material in the compost stimulates the intake of nourishment, retains moisture and encourages beneficial soil bacteria and organisms, which exist in the top layer of soil, to flourish.

Manure, which is well-rotted in the compost, is beneficial, but it is advisable not to add fresh manure unless this is mixed in the soil near the base of the hole. Proteas grow very successfully without any manure in the soil, so that it is often assumed that they should not be given any, but no detrimental results have been noticed if manure is given in moderation and if it is not of an alkaline nature. Chicken manure is more acid than cattle manure, as a rule.

The best results in growing Proteas and in the quality of their flowers are obtainable in soils containing compost, even if this is mixed with pure sand. Depth of good soil is important. The deeper and the better the soil, the more likely the plants are to thrive and have a longer life span.

ACIDITY AND ALKALINITY OF THE SOIL

On the whole, Proteas like an acid soil. Most of them grow in acid soils in nature, although a surprising number grow on limestone soils and can tolerate alkaline soils in the garden (see list overleaf).

Most garden soils are slightly acid and most gardeners in South Africa can grow Proteas without having to worry about testing their soil. Gardeners in countries like California and Israel, where soils are chiefly alkaline, have difficulty in growing Proteas successfully, although some good results have been obtained. Experiments by Dr. James P. Martin at the University of California, Riverside, revealed that best results were obtained with soils of a pH of about 5.0 to 5.5. The addition of acidifying materials such as flowers of sulphur or aluminium sulphate would help in soils which are not quite acid enough. The addition of peat-moss or even sawdust (treated with nitrogen) helps to balance an alkaline soil as well as lighten its texture. It is difficult, however, to keep the balance artificially in strongly alkaline soils and the addition of chemicals might accentuate trace element toxicity.

Many elements are readily absorbed by the Protea, as has been proved by tests and experiments. A gardener I know applied a mulch of wood-ash to four *Protea cynaroides* plants, which promptly died overnight. Knowing that the soils of the south-western Cape are frequently potash-deficient, I felt that this indicated that the plants were unaccustomed to tolerate potash. Other gardeners, however, have added woodash to their compost with no detrimental results. Potassium or potash content in the soil can reduce magnesium absorption by the Protea and cause a magnesium deficiency, which is indicated by leaves with pink to purple colouring, leaf tips curling and other nutritional symptoms. Fertilisation with magnesium nitrate in experiments by Dr. Martin improved leaf appearance, while the addition of ammonia nitrogen accentuated the magnesium deficiency.

Symptoms of nutritional troubles can best be interpreted by scientists and it is useless for the layman to add trace elements in an effort to save his plants. The slightest overdose of a trace element can kill a plant and those who advocate the free use of chemicals like Epsom-salts, for example, are taking as much chance of killing the plant as in curing it. Trace elements should only be added when it can be proved by scientific experiments that a deficiency is present.

The ordinary gardener who has failures can simply provide a good soil and favourable physical conditions, trying different species if he has no luck with any one kind. It often happens that the same species will not grow well in different positions in the same garden, so that this must always remain a matter of individual judgement and experiment.

Proteas are remarkably tolerant of different soil conditions in nature—some species more than others. *Protea obtusifolia*, for example, will grow in acid soil, but also grows plentifully in limestone formations in the Cape, so that it is one of the first Proteas that should be tried by anyone in areas or countries with alkaline soil.

Limestone veld in the Cape exists in the Bredasdorp area, starting at Stanford, about 10 miles past Hermanus near the sea and extending towards the Breede River mouth, a few miles beyond Swellendam, as well as in patches towards Mossel Bay. This is known as the Bredasdorp outcrop and the soil is alkaline and even chalky in parts.

Some Proteas which grow normally in acid soils in the south-western Cape, also grow in chalky banks near the Breede River and in other parts of this alkaline section. This seems to indicate that, while they will grow in acid soil or in ordinary garden soil, they may also be lime-tolerant. Such plants include *P. neriifolia* and *P. repens*, as well as several others that grow naturally in a wide range of different soils in nature.

The following list may be useful to gardeners in areas with alkaline soil or may aid gardeners who would like to experiment with different soil conditions. All the following Proteas and members of the Protea family have been observed growing in the limestone formations of the Bredasdorp outcrop.

LIME-TOLERANT PROTEAS

Protea pudens
P. neriifolia
P. obtusifolia
P. repens (especially white)
P. subulifolia
P. susannae

Leucospermum cuneiforme
L. cordifolium
Aulax umbellata
Leucadendron uliginosum
L. truncatum

WATERING PROTEAS

As mentioned near the beginning of this chapter, Proteas are remarkably drought-resistant. Some like more moisture than others and the amount they need can be estimated again by studying them in nature. Subjects that are difficult to cultivate, like the Marsh Rose and Stokoe's Protea, may require the cool cloud mists that bring them gentle moisture at the coast. Proteas that grow wild in areas with high rainfall, such as are found near Knysna, where the rainfall is from 50 to 100 inches annually, distributed throughout the year, are more likely to require plenty of water in the garden. Such a plant is the King Protea, *P. cynaroides*. On the other hand, it is one of the widest-spread members of the genus and also grows wild all along the Cape mountains towards Port Elizabeth and Grahamstown, where the rainfall drops to 15 or 25 inches, distributed more or less throughout the year. In the south-western Cape near Tulbagh and Worcester, *P. cynaroides* receives a purely winter rainfall, but is moistened by cloud mists in summer. The King Protea, therefore, is a versatile and easy plant to care for in the garden and can be watered freely and regularly.

Proteas and members of the family that grow on steep, well-drained mountain-sides in nature, like the Mountain Rose (*P.*

nana) and the Blushing Bride (*Serruria florida*), do not take kindly to too much watering, requiring an especially well-drained position if they are growing in the summer rainfall area, in order to offset the damping-off effects of heavy summer rains.

As a general rule, Proteas that grow wild in the winter rainfall area of the south-western Cape require regular watering during winter, whenever they are grown in gardens which receive rain chiefly in summer. A good watering twice a week will be sufficient and this can be extended to once a week or even once a month when the plants are older and well established. The frequency with which one waters depends a great deal on the type of soil that one has. Sandy soils need water more frequently than spongy, loamy soils that retain moisture. Very small plants need to be watered three times a week in dry weather as they will die if the soil is allowed to dry out completely. More small Proteas have died from being dried out than from over-watering.

Proteas which grow wild in the summer rainfall area, like *P. caffra* and *P. roupelliae*, are not such attractive horticultural subjects as those of the south-western Cape. When these are grown, however, they should be kept on the dry side in winter and watered during dry periods in summer only when rain has not fallen for two or three weeks. Again, very young plants need to be watered more frequently.

The method of watering Proteas seems to be important. It is best to water Proteas at ground level and not to use overhead watering or sprinkler systems. One loses Proteas in gardens quite often after a long period of heavy rain, especially if these rains come during summer, resulting in hot days with high humidity.

The leaves of Proteas, with their leathery texture and minute hairs, are designed to withstand the drying effect of winds which are their heritage in the mountains of the Cape and the Drakensberg. Sometimes these winds are merely breezes and they are often moisture-laden from the sea and cloudmists. They are seldom the sirocco type of hot desert wind that is desiccating and harmful, often killing small Proteas plants overnight in places like Los Angeles. Nevertheless, all wind has a drying effect and Proteas are plants which have grown up with winds, needing fresh air around them and being wind-resistant near the sea.

They do not like the cloying humidity that is more suitable

for the growth of orchids than of Proteas. As early as 1881, Sir Joseph Hooker, then Director of Kew Gardens, deplored the fact that Proteas, which had been grown successfully at Kew in former times, were no longer seen, "mainly due to the introduction of those improved systems of heating houses and that incessant watering, that favours soft-wooded plants, and is death to the Proteas of South Africa and the Banksias of Australia." It seems, therefore, that overhead watering, which dampens the leaves and does not allow them to dry out rapidly, may cause Cape Proteas to damp off as quickly as if they had had to endure 4 or 5 days of heavy Transvaal or Natal summer rains. Small plants, particularly, damp off easily from overwatering, chiefly because one waters them with a watering can and it is difficult to direct the water only on to the soil.

When watering Proteas, therefore, direct the hosepipe at soil level and give each mature plant the equivalent of at least 4 gallons of water at each watering. Commercially grown Proteas should be watered by means of irrigation ditches rather than by overhead watering. Overhead watering will also damage Protea and Pincushion flowers and shorten their lasting quality.

Several commercial growers in Natal have grown Proteas on well-drained hillsides in the vicinity of Kloof and Pietermaritzburg where humidity and heat are high in summer. They irrigate at soil level and their plants are outstanding in their sturdiness and abundance of flowering. Proteas and Pincushions have also been grown successfully about five miles from the sea at "The Valleys" near Port Shepstone, at an altitude of 750 feet, as well as on a sea front plot three miles south of Margate, where compost has been added to the pure sand. The coastal wind possibly tempers the high humidity and sub-tropical conditions.

MULCHING

Many gardeners mulch their plants in order to conserve water and keep the roots cool, using organic materials as well as stones.

In areas where wood-eating termites (white ants) occur, however, it is dangerous to mulch Proteas with materials such as fibrous compost, cottonseed husks, pine needles or sawdust, as this attracts the insects which then turn their attentions to the woody stem of the Protea. This can result in even a large bush being killed overnight. One can cover the surface of the soil with finely sifted, well-decomposed humus in order to retain moisture. As this should have been incorporated in the soil, however, it seems hardly necessary to add further dressings. Avoid adding layers of compost that might build up the soil level around the main stem of the plant.

Stone mulching is permissible in dry areas, but it is usually unsightly. Use small, rounded, decorative stones if this is desired. The stones in a rockery will achieve the effect of keeping the roots cool and this is more attractive than using a stone mulch.

Digging or cultivating on the surface of the soil to form a "dust mulch" is outmoded. This does not prevent the soil below from drying out, its only merit being that it breaks the crust that sometimes forms and prevents oxygen from entering the soil. Digging around Proteas is not advisable as they should not have their roots disturbed, especially when young.

SITUATION

An open sunny position, with at least six hours of sunshine each day, is ideal for Proteas. They can do with half a day's sunshine, but they prefer full sun. A position such as one would choose for roses is suitable for Proteas. The only exception would be in places with severe frost in winter, where it would be necessary to give the plants a warm, protected situation.

Whether Proteas like morning or afternoon sun has not been established. One can often see them growing on one side of a mountain in the Cape and not on the other side, as if a line had been drawn to demarcate their growth. The same species will sometimes grow with morning sun and sometimes with afternoon sun. They all require sunshine and grow very poorly if they receive too much shade from nearby trees.

In areas with cold winters, it would probably be best if they were shaded from the early morning sun and exposed to the hot afternoon sun, for they do not seem to wilt if they are in a hot dry situation.

Proteas also like plenty of fresh air and tolerate wind, especially cool fresh winds. For this reason they do not like a situation against a wall where air cannot circulate freely. It is sometimes necessary to give Proteas and Pincushions a warm situation in areas with cold winters. If this is the case, then they prefer a position against a dense north-facing hedge rather than against an impervious wall. A north-west corner made by two hedges meeting is ideal, as the morning sun on the leaves causes the most frost damage.

PROTECTION AGAINST FROST

Several Proteas grow above the snowline or at high altitudes in the Cape mountains and should tolerate cold well. These include:-

Protea acuminata
P. aristata
P. amplexicaulis
P. canaliculata
P. magnifica
P. cryophila
P. eximia (P. latifolia)
P. grandiceps
P. humiflora
P. lorifolia (P. macrophylla)
P. effusa

Protea pityphylla
P. punctata
P. repens
P. rubropilosa
P. venusta
P. witzenbergiana
Leucospermum catherinae
L. grandiflorum
Leucadendron album
L. arcuatum
L. rubrum

One must remember, however, that the snows of the south-western Cape come together with winter rains and, where there is moisture, frost is not severe or crippling. Snow is seldom seen in winter on the Transvaal highveld, but there are often black frosts, when the icy, crisp night air, accompanied by cold dry winds, burns the tips of tender growth and even kills plants altogether.

The Pincushion Flowers (*Leucospermum*) seem to be more tender to frost than the Proteas. Some Leucadendrons are fairly hardy, while others, like the Silver Tree (*Leucadendron argenteum*) are very frost sensitive. *Leucadendron tinctum* may endure cold in the Cape mountains, but large branches frequently die off after a cold spell on the highveld. Protea bushes and some Pincushions have survived the coldest winter nights in Johannesburg, when the night temperature dropped to below freezing point, but the flowers have been spoilt. In the case of Pincushions, secondary buds will open in spring if they have been untouched by the frost.

Frost is the limiting factor in growing Proteas, but if one can protect the plants in the young stages, then one has a reasonable chance of bringing them to the flowering stage. Once wood has formed in the stem, after about two years, then the majority of Proteas will not be killed by cold.

The best way to shelter young Protea plants from cold winter nights is to cover them each evening with an upturned cardboard box, such as one obtains from the grocery store. This will keep off the frost and insulate the plant from the cold. The box can be removed during the day, so that fresh air and sunshine can play on the plant and it will not be deprived of light. It is a mistake to cover the plants in such a way that they receive no sunlight during winter, for it will be found, on removing a permanent cover, that the plants might have died of suffocation and lack of light. If it is too much bother to perform the regular duty of covering the plants each night, one can construct a permanent cover, provided that one creates two essentials. In the first place, the covering should let in some air and be made, preferably, of long grass or of a permeable material like hessian. These materials can be placed around a wigwam of sticks, leaving space all around and above the plant. Secondly, the entire wigwam should not be covered in, but the warmest side left open, so that sunshine can reach the plant. The northern aspect is the warmest in the Southern Hemisphere. A fence of transparent polythene, leaving part of the top open, is satisfactory, provided that there is quite a lot of space around the foliage.

It will be found that one needs to protect the plants only for one or two winters. By the third season, the plants have generally grown large enough to be above ground frost and will frequently reach the flowering stage.

Severe cold will sometimes affect mature plants, burning both flowers and foliage or even killing the plant, but there is little that one can do to prevent this from happening, except to anticipate the cold and give the plants a sheltered position from the beginning.

PESTS AND DISEASES OF PROTEAS

Proteas seem to be remarkably free of pests and disease in gardens. Losses generally occur from purely physical causes such as bad drainage, incorrect pH of the soil or extreme frost, apart from sheer neglect.

Fungus diseases have been known to attack Silver Trees in the Cape Peninsula, causing the death of many of them, but there is little one can do to prevent this. Damping-off, which is a fungus disease, kills many small seedlings, but this is generally brought on by poor drainage and overwatering during hot weather and rainy conditions. Overcrowds of seedlings should be avoided and they should not be allowed to get too wet.

Scale insects, particularly those which resemble white pinheads, sometimes cover the leaves of Proteas, but the plant will live for many years without obvious harm from these and they tend to disappear in time. Do not apply oily liquids or emulsions to kill the scale, but dust with mercaptothion in preference. As ants increase scale infestation, they should be kept under control (See *The Complete Gardening Book*, Chapter 37).

CARE AFTER FLOWERING—PRUNING OF PROTEAS

Protea flowers do not drop off the bush, but close up like a bud to protect the seeds that will form before turning brown and dry. The seeds are released after a few months and the starry brown base of the flower-head remains on the plant. These brown, dry flowers make the bush look very untidy in the garden and should be removed immediately after flowering. One can either twist off the flower-head by hand or cut it back with a good length of stem.

Pruning after flowering helps to keep the Protea bush compact and promotes new healthy growth.

Removal of the flowers prevents unnecessary seed formation which uses the plant's energy, so that this is essential for the general health of the bush.

Pincushion Flowers (*Leucospermum*) do not remain on the bush after flowering, but disintegrate and drop their seeds on to the ground. Pruning of Leucospermums, therefore, is not as essential as the pruning of Proteas. Nevertheless, the flowers can be freely picked for the vase and the bush should be kept compact by cutting back lanky stems. This is particularly important in the case of *L. reflexum*, which grows up to 10 feet and flowers near the top if it is not kept in trim from the early stages.

Disbudding of flowers is not necessary with Proteas, but can be done to prevent twin or triple flowers forming on Pincushions. It is not wise to disbud Pincushion flowers too severely in areas with cold winters, however. The first early bud is frequently hardened and killed by the frost. Later, during spring, fresh buds that form beneath it will develop into perfect blooms. Once the weather has warmed up one can remove one bud if "double" flowers are seen to be developing. They are attractive and showy, however, so that it is not really necessary to disbud Pincushion flowers at all.

If a Protea or Pincushion bush has begun to die back as a

result of lack of water, frost damage, or old age, pruning can, quite often, give it a new lease of life. This should be done in the spring after flowering. Remove all dead wood and cut back long branches with dead foliage on them. Remove all dry flowerheads. Try to leave as much green foliage on the bush as possible, even if it is high up, for to remove green foliage altogether might cause the death of the whole bush. After pruning, water well and place a dressing of manure or compost mixed with a little fertilizer, on the surface of the soil. New growth is usually most prolific in spring and the plant will soon be clothed with young leaves.

GROWING PROTEAS IN A NUTSHELL

A few golden rules on growing members of the Protea family may be distilled from the above chapter.

1. Proteas require well-drained soil of good depth, preferably of a gravelly or sandy nature, with compost added. Sloping ground is helpful in places where soil is not well-drained or rainfall is heavy.
2. Most Proteas require an acid soil, but there are some that are lime-tolerant.
3. Regular watering should be given, especially during winter. Proteas are drought-resistant especially near the coast during summer, but will tolerate a heavy summer rainfall more readily if the soil is well-drained. Do not water with overhead irrigation which may cause damping off.
4. Proteas require fresh air and are wind-resistant, but dislike hot, dry desert winds, preferring cool breezes.
5. Proteas thrive best if they have at least six hours of strong sunshine each day.
6. Protect members of the Protea family from severe frost. Proteas are hardier than Pincushion Flowers (*Leucospermum*) and the Silver Tree (*Leucadendron argenteum*), but some species are hardier than others.
7. Remove Protea flowers immediately after blooming or pick them freely for the vase. Prune back long stems on Pincushion Flowers to keep the bushes compact.

CHAPTER V

CULTIVATION OF PROTEAS AROUND THE WORLD

Proteas have caught the imagination of gardeners and commercial growers throughout the world. Protea growing is in its infancy even in South Africa, while the fantastic possibilities of its importance as a commercial cut flower have only been surmised. Growers can scarcely keep up with the demand and the future of Proteas both in the garden and at the florist seems very bright.

The country which has been most successful with Protea cultivation is New Zealand, where several prominent nurserymen have grown plants on a large scale. The climate and soil are excellent and Proteas grow there without many problems. They also do well in Australia, particularly in the Melbourne area, and on the island of Maui, Hawaii.

Proteas have been tried by enthusiasts in California, where the climate resembles that of the Cape of Good Hope, with its winter rainfall and long, hot summer. Unfortunately, the soil is chiefly alkaline and the rainfall is sparse, so that there are natural difficulties to overcome. Nevertheless, several growers have been succesful and, no doubt, Proteas will be more popularly grown when more is understood about their requirements. Their cultivation has proved successful near San Francisco and on a large scale at Escondido, near San Diego.

Proteas have also been raised from seed in Israel—another country with a similar climate to that of the Cape. They have not been very successful owing to the chiefly alkaline soil, but this problem may be overcome.

Beautiful Proteas have been grown in the south of France near Nice. There is a new venture started by a South African, which is to cultivate Proteas commercially in Corsica and the south of Sardinia, where the soil has been tested and found to be neutral to very slightly acid. Proteas have also been grown successfully in Madeira.

Numerous Proteas and members of the Protea family have been grown at Tresco Abbey on the Scilly Isles, Cornwall, where frost is seldom experienced. They thrive on dry, stony, acid soil, open to the sun and wind. They have not been planted in the open in Britain, but have only been grown in containers that have been moved under cover during winter.

Individual plants have been grown in drums in the greenhouses of Europe, particularly in the 18th and 19th centuries, when Proteas were regarded as "the handsomest of plants, whether for size, form or colour of inflorescence; and would carry away the first prize at any horticultural show...", to quote Sir Joseph Hooker in Curtis's Botanical Magazine of 1881. He prophesied then that they would return to popularity, which they had lost through lack of knowledge about their cultivation and again become "the wonders of the shows".

The wheel appears to have come full circle and the Protea family seems now to be coming into its own.

CHAPTER VI

PROPAGATION OF PROTEAS

With the increase in popularity of Proteas, both as cut flowers and as garden plants, the matter of propagation is becoming an important subject for review and experiment.

Up to the present time, it has been customary to propagate members of the Protea family from seed. It has been easier for the professional nurseryman to grow Proteas from seed than by any other means. Yet it is not always easy to obtain sufficient seed of some common species, while rare species are often extremely difficult, if not impossible to acquire, even in the wild.

The obvious method to consider is that of growing members of the Protea family from cuttings and, while this is not easy, it is practical, especially with the modern aid of hormones and mist spray equipment. Propagation by cuttings would ensure that Proteas could become an important commercial crop and that the flowers would be uniform in type. It would also be possible to perpetuate beautiful hybrids that are produced either as a result of natural cross-pollination or by hand-pollination.

The amateur grower is well-advised to buy established Protea plants from a nursery as it is not easy to care for the young seedlings in the first few months, even if germination is good. They do not transplant easily and damp off frequently, so that losses are generally considerable in the early stages of growth. In addition, one can save from six to twelve months by purchasing a well-grown plant that is about six or eight inches high.

I. GROWING PROTEAS FROM SEED

It is thought that Protea plants that are grown from seed and grown in the same area do better than plants which are brought in from another climatic area. Provided that the plants are healthy and have not suffered any set-back through travelling, however, there is no real proof that imported plants do badly. Excellent results are obtainable both from plants imported from another climatic area as well from those which are grown in one's own area. Seed which is gathered from plants grown in one's area, however, often germinate better and in a shorter time than imported seed.

For the gardener with patience, and certainly for the nurseryman, the cultivation of Proteas from seed is rewarding and fascinating.

TYPES OF SEEDS

The seeds of Proteas vary in each species and the seeds of the Protea family are vastly different from one another. A glance at page 221 of this book will show a selection of the various seeds that are typical and often quite beautiful.

Most Protea seeds have hairs around them so that they can be scattered by the wind and frequently resemble shuttlecocks. Some *Leucadendron* seeds have little parachute devices to carry them to their destination on the breeze, notably those of *L. rubrum*. Others, like *L. platyspermum* have disc-like scales, shaped like butterflies, to disperse them on the wind. Pincushion Flowers *(Leucospermum)* drop their smooth beetle-like seeds, likened to lice ("luisies") near the mother plant.

Many experiments have been made in trying to improve the germination of Protea seeds, such as burning or pulling off the fuzz or soaking them, but best results are generally found with untreated seed. The knowledge that natural toxins form in the husk of the seeds, inhibiting germination, has led some growers to place the seed in running water for a day or two before sowing it. There is no proof that this solves the problem of shy germination, while Protea seeds germinate so readily in well-drained soil, which in itself serves the purpose of washing away any toxins, that this pre-washing seems to be of academic interest only.

FERTILE SEED

One must realise, however, that not all Protea seed is fertile. Fertile seed can generally be recognised as it is fatter than infertile seed, but its appearance is not infallible. In the case of *Protea repens*, both fertile and infertile seeds are of almost uniform thickness, while fertile seed of *P. magnifica* is the size of a pea, so that it is easily distinguished from the infertile thin seeds.

The ordinary gardener should sow all the seed gathered from a Protea head and be prepared to give it the extra space in the hope of including all the fertile seeds.

Some Protea heads contain hundreds of seeds and the number of fertile seeds amongst them is often considerable. A well-known commercial Protea grower, who has made records of the number of fertile seeds in a Protea head, has found that a head of *Protea cynaroides* has yielded up to 400 fertile seeds, while *P. magnifica* averages 24 seeds per head, *P. neriifolia* yields 60 seeds per head and *P. repens* about 40 seeds per head. The best advice, therefore, would be to plant all the seeds available.

Protea seed takes several months to mature and is really at its best a full year after flowering. When the flower has bloomed, it will be seen to close up like a bud in order to keep out rain which will spoil the setting of seed. The long dry summer months at the Cape are ideal for the maturing of Protea seed, but seed will also mature successfully in other climates. Insects spoil quite a lot of seed in the wild.

The mature seed is generally ready for gathering when the new crop of flowers comes out. It takes several months to become brown and hard.

Protea seed can be sown about 4 to 6 months after the flowers have bloomed, but then it should be planted immediately, for the soft seed goes off rapidly. Mature Protea seed is viable even after 3 or 4 years. The age of the seed can be seen by its position on the unpruned plant, as the old seed heads do not drop off. The newer flowers open above the old ones each year. This is not the case with *Leucospermum*, as the Pincushion Flowers shrivel soon after blooming and fall to the ground, when the seed is already mature.

Seed of *Leucospermum* and *Leucadendron* sometimes matures so easily that it will ripen in the vase when the flowers are allowed to dry out. One seldom keeps Pincushions after they fade as they do not make very attractive dried material, but if one does keep them long enough, one may notice the smooth seeds protrude slowly until they can then be picked out by hand. This generally happens with the smaller species like *Leucospermum oleifolium*. Much depends on the state of maturity of the flower when it is picked as well as on the species.

The Tolbos, *Leucadendron rubrum*, is so tenacious of life, however, that the closed female flower-head will develop its open woody structure in a dried arrangement in the vase and release its seeds about two months after it is picked. The fascinating little parachutes can then be lifted out of their sockets and retained until the time is ripe for planting in the autumn.

The seed of Proteas and Pincushions is designed to withstand trying seasons in nature and it germinates erratically in cultivation. The different species are best planted in separate tins or pots, or in seedbeds which can be left undisturbed for a year or more. Some seed in a batch will germinate in from 5 to 8 weeks and the remainder may germinate a year later, so that one should not toss away the seed-tray in a hurry.

THE TIME TO SOW SEED OF PROTEAS

The best germination takes place during low temperatures at the end of summer and in autumn, from March to May in the southern hemisphere. The end of March is the best time to sow seed in the winter rainfall area of the Cape. Autumn sowing gives best results in the summer rainfall areas of the Transvaal and Natal and seed may be sown all through winter where the cold is not too severe. In places with cold winters, it may be best to sow the seed towards the end of winter, about July, so that the young seedlings do not have to contend with cold weather, but are ready to be hardened off when the danger of frost has passed.

HOW TO PREPARE THE SOIL

The soil for sowing seed of the Protea family should be light and well-drained. It can be sandy or loamy, but it should be mixed with sifted, well-rotted compost. Drainage should be excellent and the water should sink down immediately it is poured on to the surface. An underlay of coarse rubble is ideal in a large seedbed. Individual containers should have broken crocks, small stones or coarse sand at the base.

Some growers prefer to sow two or three seeds directly into individual containers, using small terracotta pots, tins or black polythene bags, in order to avoid pricking out and to make transplanting easier.

The situation for the seedbed or tins should be open and sunny. At first the seed should be shaded with hessian, straw, rushes, wooden slats or even glass, but this must be removed later so that the young seedlings can be accustomed to full sunshine.

Sow the seed on the surface of the soil and cover it with a light covering of soil, not thicker than the width of the seed or more than $\frac{1}{4}$ inch deep. Choose a windless day so that the seed does not blow away. Water carefully so that the seed does not wash away.

It is important not to allow the soil to dry out until the seed has germinated. One dry spell will stop germination completely. Water the seeds in the morning and in the afternoon in order to keep the soil damp.

Start accustoming the young seedlings to the light when 3 leaves appear and keep them partially shaded until they are about 2 inches high. Try to keep the soil damp without wetting the young leaves, especially during summer and at the end of the day. Water during the morning so that the leaves dry off before the late afternoon in order to avoid damping off and fungus diseases.

Prick out the young plants when 2 real leaves have formed or even when 6 leaves have formed, but pricking out becomes riskier the longer it is delayed. Losses often take place after pricking out and this may be avoided by planting the seed in individual containers. When pricking out, try not to disturb the mass of roots, but the tap root may be cut short if it is too long for the tin. Do not plant the seedlings into tins that are smaller than 2 lb. jam tins, while deeper tins are preferable. A 2-inch high seedling frequently has a root that is 4 to 6 inches long.

When the plants have to remain in their tins for a long time, sever the tap root in advance when pricking out, in order to encourage bushy growth. Sever any roots that grow through the tin and keep the soil damp, watering daily.

TRANSPLANTING INTO THE OPEN GROUND

Do not transplant any Protea plant into the garden if it is smaller than 4 inches. Very small seedlings transplanted to the open ground seem to "stand still". They can be transplanted when they are 6 to 12 inches in height or even larger. In the winter rainfall area of the Cape, six-month-old plants may be transplanted from March to September, before the hot, dry summer sets in. In the dry interior districts, one should plant out one-year-old to eighteen-month-old plants in spring, around September, when all danger of frost has passed. In mild winter areas, they can be transplanted as early as August. Where watering may not be regular, best results are obtained by transplanting from November to February, when summer rains are fairly regular.

Before transplanting a Protea plant from a tin, see that the soil is damp. Water it several hours before opening the tin. Cut down the sides of the tin with a sharp tip snips, so as not to disturb the ball of roots. Make three openings so that the sides of the tin can be bent back and the ball of soil lifted out intact. Some people prefer to cut off the bottom of the tin and plant the whole tin in the ground. This will rot in time. It is easy to tip out a Protea plant growing in a terracotta pot by knocking the rim of the pot, upside-down. Certainly the easiest method of freeing the plant is to cut away a plastic bag.

Do not plant the Protea or Pincushion any deeper than it was growing in the tin, or the stem may rot. High planting is advisable where drainage is poor. Firm the new soil around the plant and water it immediately to settle it.

Protea plants are not very delicate and they may be handled differently depending on the climate of the area in which they are grown. Protea seedlings grown by a well-known New Zealand nurseryman, are grown in rows in the open ground until they are as large as two or three feet in height. Their roots are trimmed a few weeks before they are required, by driving a spade into the soil on either side of the plant and severing the tap root below. These plants are watered and settled down again. They are dug out a few weeks later, their roots and enclosing soil "balled" or wrapped firmly in hessian and transferred to their position in the garden without loss. The moisture in the air enables one to handle the plants in this way, but the method is not recommended in a dry climate.

HOW LONG DO PROTEAS TAKE TO FLOWER?

Proteas generally flower in the third year from seed, but more commonly in the fourth year, especially in cooler climates. Where conditions are extremely favourable, they may bloom in the second season. This varies according to the species and generally happens with those which are easiest to grow like *Protea burchellii, P. eximia, P. cynaroides* and *P. obtusifolia*. Some of these will flower as early as 18 months after sowing the seed.

Pincushion Flowers generally flower a year later than Proteas, but there are always exceptions. *Leucospermum lineare* has produced a flower in the second season in a warm position, yet *L. reflexum* usually flowers in the fourth season. The Blushing Bride (*Serruria florida*) often flowers after a year if conditions are favourable.

II GROWING PROTEAS FROM CUTTINGS

The advantages of growing Proteas and Pincushions from cuttings are obvious. To be able to propagate the colour forms of

Leucospermum cordifolium, that has a colour range from tangerine to pale pink tones, is an exciting possibility that would make these plants even more desirable than they are. Natural and hand-pollinated hybrids that may or may not come true from seed, or may not even set fertile seed if there are no sun-birds or suitable insects to pollinate them, could be retained for posterity if they could be grown from cuttings. When growing Proteas commercially, it would be more practical, quicker and more dependable to grow them from cuttings than by any other means.

Success in growing Proteas from cuttings depends a great deal on favourable climatic conditions. The original *Protea repens* grown by another New Zealand nurseryman was grown from seed, but scores of plants were then raised from cuttings of these bushes with success. Leucadendrons and Proteas were also grown from cuttings and found to root easily in March and April.

Growing Proteas from cuttings is not easy. In a book on *Proteaceae,* published in 1809 by Joseph Knight, gardener to the famous George Hibbert of Clapham, who had the largest collection of living Proteas in Europe in his famous garden, he states that growing Proteaceous plants from cuttings is difficult. He was successful in growing them from seed and noted then that autumn and winter were the best seasons for germination, since the seeds germinated best at low temperatures.

The introduction of mist spray equipment in modern times, however, has made it possible to root members of the Protea family with comparative ease.

The idea behind using mist spray is that the moisture keeps the cuttings fresh until they form roots. Its use also helps during ordinary transplanting on hot days and prevents losses through drying out.

The mist spray for Proteas must be used in the full sun, for the sunshine makes food in the leaves and forces root growth. The mist should only be switched on when the sun is bright and hot and must be switched off at night and on dull, cloudy days or the cuttings will damp off. The best time to grow cuttings under mist spray conditions is, therefore, during summer, when the sun is warm and bright. As cuttings generally "take" best in spring and autumn, the best time would seem to be from mid-September until November, but one could go on taking them at intervals until the autumn.

Tip cuttings of from 8 to 10 inches long are best. They should not be very soft-stemmed, but taken with firm or harder wood. It is necessary to keep as much foliage on the Protea cuttings as possible. Strip off the lower leaves and bury the cutting for half its length, retaining as many leaves as possible above the surface. Hormone powders may be used before inserting the cuttings into the sand, if desired.

For large-scale or commercial use, it is best to construct a special frame for the cuttings. Build a 2-foot deep brick frame with a sloping cement base, so that the water can run out rapidly through weep holes made at the base of the bricks. The water must not stand, but run off rapidly. Place a layer of coarse rubble on the cement, covered with finer rubble and, finally, fill up with pure coarse sand. There should be no nourishment in the sand at all, and no feeding is given.

The cuttings of Proteas, Leucospermums, Leucadendrons, Mimetes, etc., take at least three months to form roots under mist. The Protea cuttings can be tinned up into individual containers of good soil as soon as the roots develop. It would be wise to keep the transplanted cuttings near the mist for several weeks until the transplanting is proved to be successful.

Members of the Protea family have been grown from cuttings under mist by growers in the Cape, Natal and the Transvaal. On the whole, these plants have not survived long after being transplanted, but this may be due to lack of interest and effort. This method is well worth investigating further and organising efficiently. In view of the fact that the method is used successfully in New Zealand, there seems to be no valid reason why it cannot be used in this country. A member of the Australian Proteaceae, the *Macadamia* nut, has been grown from cuttings by the thousand in Kenya, at a Government Experimental station outside Thika, and vigorous trees have resulted from slipped plants. There may be no horticultural connection between these genera, but the indications are that growing cuttings by means of mist-spray cultivation needs to be tried out methodically on a large scale.

Since first publication of this book, commercial growers have begun to propagate Proteas by means of cuttings on a large scale.

CHAPTER VII

CHOOSING PROTEAS

Gardeners are the people who are interested in choosing Proteas. As they are also interested in picking the flowers for the vase, the obvious way in which to select Proteas is to judge the beauty of their flowers as well as the floriferousness of the bush.

Were a gardener to have space for only two or three members of the Protea family, it follows that one would recommend those plants with the loveliest flowers, taking into account whether they are easy or difficult to grow and whether they are obtainable in the trade. Bearing these considerations in mind, the author has made a selection of choice members of the Protea family in order to guide the novice who wishes to start the nucleus of a collection.

All Proteas are not beautiful and some have quite ragged flowers. Many are interesting botanically while some are merely curious. *Protea cordata* and *P. cryophila*, with their strange, short-stalked flowers at ground level, intrigue the collector or the botanically minded person, but are never likely to become popular garden subjects. Many people like to grow plants because they are unusual and there are certainly many unusual specimens in the Protea family. In order to reflect a true picture of the Protea family, many of these unusual plants have been illustrated in this book and, although one is fascinated by their differences, one would not necessarily wish to grow them. Some have less obvious attractions that grow on one with familiarity. *Protea amplexicaulis*, for example, has purplish black flowers that grow with their heads turned towards the ground so that they scarcely enhance the plant. The long, trailing stems of attractive foliage, however, can be decorative in a vase and one could grow this plant only for its foliage. If one took the trouble to pick the flowers, one could then make an unusual arrangement.

There are many unusual members of the Protea family that make interesting and even exciting flower arrangements, for the more unusual they are, the more one's attention is arrested by them in the vase. The choice below, however, is based purely on horticultural merit, with the emphasis on those species which are most spectacular in the garden and have the loveliest flowers.

It is not easy to group Proteas according to colours without repetition, as many of the attractive kinds come in several colour forms. Proteas are generally available in shades of pink, ranging from palest blush pink to deepest rose or pomegranate red. Some are available in white, cream and yellow and others are greenish. Some of the colours are extremely subtle and difficult to fit into any category, but the list below will serve as a rough guide in choosing a range of tones. The Leucospermums drift mainly into the orange and flame tones, while the foliage of many Leucadendrons turns yellow or russet in the spring.

One of the loveliest members of the Protea family, the Marsh Rose (*Orothamnus zeyheri*) is conspicuous by its absence in the following lists, as it is rare and unobtainable, besides being difficult to cultivate and a protected wild flower.

THE PICK OF PROTEAS

FIRST CHOICE OF PROTEAS

THE TEN BEST PROTEAS THAT ARE FREELY AVAILABLE

Protea magnifica. Giant Woolly-beard. Rose or red with a yellow or cream variety. Not the easiest to grow.

Protea compacta. Bot River Protea. Clear pink flowers. Early bloomer. Easy to grow.

Protea cynaroides. King Protea. Pale pink or deep red flowers. Also a cream variety. Huge flower on a small bush. Easy to grow. Adaptable to soil conditions.

Protea grandiceps. Peach Protea. Magnificent peach pink or reddish flowers. Fairly slow.

Protea laurifolia. Fringed Protea. Pale pink flower with brown and white furry tips. Easy to grow.

P. repens. True Sugar-bush. Ware Suikerbos. Pink, red or white flowers. Long sticky bud. Easy to grow.

P. neriifolia. Oleander-leaved Protea. Salmon or deep rose flowers, tipped with black fur. Free blooming and easy to grow.

P. speciosa. Brown-bearded Protea. Pink bracts tipped with brown fur, excellent in the vase.

P. burchellii. Gleaming Protea. Pale rose or ruby red, with attractive sheen. Blooms prolifically.

P. obtusifolia. Greenish bracts shading into deep red tips. Grows in alkaline soil.

SECOND CHOICE—
AN ADDITIONAL TEN PROTEAS

Protea aristata. Ladismith Protea. Deep rose pink flowers. Soft needle-like foliage. Summer-flowering and hardy.

P. lacticolor. Baby Protea. Small, clear pink. Wide-open flowers. Large bush.

P. longifolia. Long-leaved Protea. Soft white flowers with black centre, sometimes pale pink.

P. lepidocarpodendron. Black-bearded Protea. Dramatic black-tipped creamy flowers.

P. stokoei. Stokoe's Protea. Beautiful pink flowers, but rare.

P. nana. Mountain Rose. Wine coloured, small drooping flowers.

P. pudens. Aardroos. Charming small flowers, creamy, tinged rose. Not easy to grow.

P. pityphylla. Pine-leaved Protea. Dark red flowers. Rare and not easy to grow.

P. susannae. Susan's Protea. Pink, tinged brown flowers. Pungent foliage. Grows on limestone.

P. coronata. Apple Green Protea. Apple green flowers. Mixes well in the vase.

FIRST CHOICE OF
TWELVE OTHER MEMBERS OF PROTEA FAMILY

Leucospermum cordifolium. Nodding Pincushion. Flame, pink, yellow or orange flowers. Prolific bloomer. Excellent in landscape.

Leucospermum reflexum. Rocket Pincushion. Salmon-crimson flowers. Prolific bloomer. For large gardens.

Leucospermum tottum. Fire-wheel Pincushion. Soft pinkish flowers. Blooms later than others.

Leucospermum catherinae. Catherine's Pincushion. Unusual champagne yellow, large flowers.

Leucospermum lineare. Narrow-leaf Pincushion. Flame or orange flowers, attractive foliage.

Leucospermum vestitum. Upright Pincushion. Orange-yellow, large flowers.

Leucospermum cuneiforme. Yellow flowers aging red. Prolific and decorative in landscape.

Leucadendron argenteum. Silver Tree. Decorative silvery foliage.

Leucadendron discolor. Sunshine Bush. Large golden bracts with red centres (male). Prolific and showy in landscape.

Serruria florida. Blushing Bride. White, flushed pink flowers. Dainty and long-lasting.
Mimetes cucullatus. Red Bottlebrush. Rooistompie. Yellow, tipped red leaves. Showy in garden and vase.
Aulax umbellata. Yellow "feather-duster" flowers. Showy in landscape.

SECOND CHOICE OF
TEN OTHER MEMBERS OF PROTEA FAMILY

Leucospermum rodolentum. Grey foliage, yellow flowers. Small.
Leucospermum conocarpodendrum. Kreupelhout. Bright yellow flowers half hidden in foliage. Wind-resistant.
Leucospermum oleifolium. Tufted Pincushion. Small yellow flowers aging red. Small bush.
Leucospermum grandiflorum. Rainbow Pincushion. Yellow flowers, red-tipped, aging red. Colourful.
Leucospermum muirii. Small orange-yellow flowers on small bush.
Leucospermum prostratum. Creeping Pincushion. Decorative small yellow flowers, deepening to apricot and red as they age. Difficult to grow. Rare. Scented.
Leucadendron xanthoconus. Geelbos. Gold-Tips. Male plants have very showy yellow leaves in spring. Showy.
Leucadendron tinctum. Rose Cockade. Top leaves flushed red in spring. Showy.
Paranomus spicatus. Perdebos. Mauvish-pink flower-spikes in spring.
Paranomus reflexus. Green Bottlebrush. Yellowish-green flower-spikes which are curious in the vase.

CLASSIFICATION OF COLOURS IN THE PROTEA FAMILY

Pink

Protea magnifica	*Protea burchellii*
P. compacta	*P. punctata*
P. caffra	*P. repens*
P. cynaroides	*P. roupelliae*
P. lacticolor	*P. speciosa*
P. laurifolia	*P. stokoei*
P. mundii	*P. susannae*
P. neriifolia	*Leucospermum tottum*

Pale, flushed pink
Protea pudens
Paranomus spicatus
Serruria florida
S. barbigera

Strong Red or Rose
Protea aristata
P. magnifica
P. cynaroides
P. grandiceps
P. obtusifolia
P. burchellii
Protea repens
P. eximia
P. laetans
Leucospermum reflexum
L. gerrardii
Mimetes cucullatus

Wine Red
Protea acuminata
P. effusa
Protea nana
P. pityphylla

Flame and Orange Tones
Leucospermum cuneiforme
L. oleifolium
L. vestitum
L. grandiflorum
L. lineare
Leucospermum muirii
L. cordifolium
L. prostratum
Leucadendron arcuatum
Mimetes hirtus

Yellow and Cream
Protea nitida
P. magnifica
P. lepidocarpodendron
P. obtusifolia
L. catherinae
L. rodolentum
L. conocarpodendrum
Leucospermum cordifolium
Leucadendron daphnoides
L. discolor
L. eucalyptifolium
L. xanthoconus
Aulax cancellata
Aulax umbellata

Green
Protea acaulos
P. coronata
Protea scolymocephala
Paranomus reflexus

White
Protea longifolia
P. aurea
Protea repens
Leucospermum bolusii

PROTEAS THROUGHOUT THE YEAR

While it is impossible to state categorically when Proteas may be expected to flower, the following groups will indicate the times when they are usually found in bloom and help the gardener to make a selection so as to have Proteas almost throughout the year. Some Proteas flower intermittently for nine months of the year and the great majority flower in the spring. They are valuable stand-bys for autumn and winter, when there are so few flowers in bloom, and some Proteas are available in midsummer. The only time when Proteas are scarce is in February and the beginning of March, before the new season's flowers have started. Even then, a few like *P. lacticolor* and *P. mundii* may be in bloom. The following species have been selected from those which make the best cut flowers or are grown most frequently.

Some Proteas for Autumn and Winter Flowering

Protea nitida
P. cynaroides
P. compacta
P. eximia
P. lacticolor
P. lepidocarpodendron
P. longifolia
P. coronata
P. pudens
P. mundii
Protea neriifolia
P. obtusifolia
P. punctata

Protea repens
P. scolymocephala
P. speciosa
P. stokoei
P. susannae
Leucospermum cordifolium
L. vestitum
Paranomus reflexus
Leucadendron salignum
L. xanthoconus
Serruria aemula
S. florida

Some Proteas for Spring Flowering

Protea magnifica
P. acuminata
P. compacta
P. cynaroides
P. eximia
P. grandiceps
P. lacticolor
P. laurifolia
P. lepidocarpodendron
P. longifolia
P. coronata
P. pudens
P. nana

Leucospermum bolusii
L. cordifolium
L. cuneiforme
L. catherinae
L. conocarpodendrum
L. oleifolium
L. grandiflorum
L. vestitum
L. lineare
L. muirii
L. reflexum
Serruria pedunculata

Protea obtusifolia *Serruria barbigera*
P. burchellii *Mimetes cucullatus*
P. repens *M. hirtus*
P. rubropilosa *Paranomus reflexus*
P. scolymocephala *Paranomus spicatus*
P. susannae *Aulax umbellata*
Leucadendron, most species *A. cancellata*

Some Proteas for Midsummer Flowering

Protea aristata *Leucospermum catherinae*
P. caffra *L. cordifolium*
P. cynaroides *L. conocarpodendrum*
P. eximia *L. oleifolium*
P. grandiceps *L. vestitum*
P. aurea *L. tottum*
P. mundii *Mimetes cucullatus*
P. roupelliae *M. hirtus*
P. scolymocephala
P. speciosa
Leucadendron discolor

CHAPTER VIII

LANDSCAPE DESIGN WITH PROTEAS

When one is growing Proteas to begin with, one thinks only of how to grow the plant and seldom of its appearance in the landscape. On most occasions, a situation is chosen for its climatic and soil conditions, without regard for the ultimate appearance of the plant.

Too many Protea enthusiasts are apt to plant Proteas haphazardly in lawns, relying on the beauty of the plants to create interest, without really providing an attractive setting for them. The final effect of such a garden is more like that of a collection, which sometimes has a ragged appearance.

If one has a good position for Proteas, such as a north-facing gentle slope, one can plant a collection together in order to facilitate treatment, but one should aim at some design in placing the plants. There is a certain sameness about some Proteas, especially the larger bushes, although the colours of their leaves vary and one can avoid monotony if several are planted together. Size is of great importance and one should always place the taller growers in the background with the shorter types in front. Find out how wide the bushes spread and plan to allow them to touch when mature, but do not crowd them as this ultimately inhibits their growth and spoils their appearance.

Proteas can be used formally as well as informally, as they are evergreen and can be neatly shaped if necessary. On the whole, they are rugged and bushy, so that they are particularly successful in an informal lay-out with flowing lines and asymmetrical planting. A large herbaceous border with large Protea bushes in the background and shorter types at the side, is an ideal setting for them. A shrub border without annuals and perennials is even more successful. Try not to have more than two "rows" of shrubs without a pathway between, or it becomes uncomfortable to approach the plants in order to cut the flowers and trim them, and also creates an impenetrable jungle. Stepping-stone paths of slate can be laid between the bushes if a large area of ground is planted. For ideas on lay-out see sketches on page 56 and, for an approximate guide to the sizes of the plants, see the lists at the end of the chapter.

Proteas take about three years to fill out and reach the flowering stage, and, while their foliage is decorative and evergreen, there can be a dull patch for a while especially if they are planted in a group. One should grow a low groundcover between them which will appear colourful and yet not swamp the small plants. Grass is satisfactory, but it does require mowing and this may lead to damage to the stems by the lawn-mower. In addition, grass needs constant feeding and spreads so rapidly that it is difficult to eradicate without damaging the roots of the Protea plants.

One of the most attractive groundcovers which has proved successful while growing Proteas is the indigenous perennial *Arctotheca calendula*. This grows wild on the coastal sands at the Cape near Port Elizabeth, as well as inland, and has proved to be an excellent garden plant in recent years, thriving in rich soil or poor. It spreads rapidly over the surface, on sloping banks or level ground, and produces large yellow daisy flowers throughout most of the summer months. Although normally evergreen in mild climates, it is temporarily deciduous if frosts are severe, but usually retains its leaves under the shadow of the bushes. It likes full sunshine, as do Proteas, but will also thrive in partial shade. It is drought-resistant enough so as to make the watering of the Protea plants of greater importance.

Another excellent type of perennial groundcover for Proteas is the *Lampranthus* or any other member of the *Mesembryanthemum* family. These plants are bright with flowers in the spring and cover the ground rapidly with their succulent foliage, needing very little water. They usually need to be replaced by new cuttings after three years, so that they can be rooted out to good purpose as the Proteas grow to maturity. Other suitable groundcovers are *Cerastium* and Bronze *Ajuga reptans*.

Low-growing annuals can be used to make colourful patches between Proteas, but be careful not to let them smother young plants and choose types that do not need constant watering. Suitable sun-loving annuals that will flower in spring are *Dorotheanthus*, *Cotula barbata*, Virginian Stock and Violas. Violas will continue to bloom all through summer, but the others can be replaced with summer annuals such as Portulaca, dwarf Ageratum and Alyssum.

Perennials which do not spread rapidly, such as Alpine Phlox, tufted Gazanias and *Gerbera jamesonii*, can be used, but they will need to be moved as the Proteas grow and may not survive.

Bulbs are unsuitable between Proteas and there is a danger of damaging the Protea roots when lifting them.

One should make more use of Proteas as plants in the general landscape than is generally done. One is apt to ignore the possibilities of their decorative quality as plants, thinking mainly of the value of their flowers. The fact that they are evergreen makes them especially desirable in the garden.

The shapes of the bushes can be very useful. The large bushy types have great value as screens to block out an unsightly spot, to create privacy and to divide one portion of the garden from another. They can be most useful, for example, in screening an orchard and, at the same time, provide an evergreen background to a colourful flower border.

Lower spreading kinds like *Protea magnifica* and *Leucospermum cordifolium* will cover the ground on a gentle slope without the use of rocks or grassy banks to hold the soil.

Rockeries provide an excellent setting for all Proteas, particularly the spreading or sprawling types and the smaller daintier kinds. Large Proteas are suitable only in a large rockery. Leucospermums are extremely attractive amongst the rocks and their habit of growth is compatible with the shape of a rockery bed. Plants like *Leucospermum prostratum* and *Protea amplexicaulis* seem to be thoroughly at home in a rockery setting, their graceful foliage overhanging large rocks. Proteas often benefit culturally if one places them in a rockery. Plants that require somewhat drier soil situations like *P. nana*, and damp off in an ordinary bed, are more successful in rockeries where good drainage offsets the effects of heavy rainfall. Some dainty and precious plants like *Serruria florida* and *P. aristata*, that may be damaged by careless handling, are protected by enclosing rocks and may even be sheltered from frost by the rocks.

Some Proteas make excellent accent plants at the tops of steps, in corners and at the edges of pathways. There are very few with narrow shapes, but *Leucadendron argenteum* makes an ideal accent where the climate is not too cold. It is decorative while small, as well as when it is fully grown. *Leucospermum reflexum* makes a spectacular accent wherever it is placed, but requires at least 10 feet in which to spread. One could use *L. cordifolium* to good effect in smaller gardens. Smaller plants with very showy flowers like *Protea cynaroides* look well in any position, but this King Protea can be attractive accentuating a focal point in the garden. It is lovely enough to stand on its own or to be mixed with other plants. Small rounded bushes like *P. nana* are neat enough to form accent plants and would be suitable flanking the top of a flight of steps.

Grey foliage is excellent in the garden as a contrast to duller evergreens. A whole portion of a large garden can be devoted to a grey and silver planting, highlighted with pink or yellow flowers. The Protea family provides at least seven plants with grey or silver foliage which could be featured on their own or in juxtaposition with other plants. Most of the Protea family has greyish-green or soft, bluish-green foliage, but the seven plants listed at the end of the chapter are distinctive in the silveriness of their leaves.

The slender needle-shaped leaves of several members of the Protea family, listed below, make them valuable from a decorative point of view. Some would make graceful tub-plants. *Leucospermum lineare* is a delightful bush with glorious flowers like *L. cordifolium,* but provides a relief from the usual broad leaves of the latter.

In gardens with difficult situations, where soil is poor or windy conditions prevail, Proteas can come to the rescue. Most Proteas are wind-resistant, for they come chiefly from the S.W. Cape where the coastal winds sometimes develop into gales. No plant looks more attractive in the wind than the silvery *Leucadendron argenteum*, which is almost lifeless without a breeze to make it sparkle. Very hot dry winds are not good for Proteas, which, in California, have been known to "look as though a blow-torch had been turned on them" to quote the words of Dr. W. S. Stewart, Department Director of the Los Angeles State and Country Arboretum.

There are always breezes blowing in the mountains all over the country where Proteas are found growing wild, so that they can withstand most windy weather, particularly if the wind is cool and fairly moist.

It has never ceased to surprise me to see how Proteas thrive in poor soil, although they do very well in good rich soil. Provided that there is good depth of soil, Proteas thrive with very little care in gravelly soil, even though it has the quality of subsoil. They are also fairly drought-resistant and frequently survive with little water beyond that of the rainfall. It seems, therefore, that Proteas can perform a useful function in the landscape by fitting difficult growth conditions.

There is no garden too small for Proteas and one can make a

choice for everyone from the attractive ones available. The smaller and daintier Proteas will grow in tubs or large pots, and could be a feature of a sunny patio or terrace. Not only will they look attractive, but they will possibly grow more successfully in places where soil conditions are difficult, for the soil in the tubs can be carefully mixed and watering more controlled.

As is the case with most landscape design, one needs imagination and knowledge of the plant material in order to use it boldly and artistically, making the most of the interesting possibilities that are inherent in each plant.

GUIDE TO CHOOSING PROTEAS FOR THEIR DECORATIVE QUALITIES IN THE GARDEN

These lists have been made from members of the Protea family that can be grown without great difficulty in gardens and from those which are likely to be cultivated. It is useless to list plants that are rare, protected and unobtainable, like the Marsh Rose (*Orothamnus*) or plants which are not likely to become popular horticultural subjects, like *Faurea*. Some of the plants listed are not yet obtainable in the trade, but are likely to be introduced if there is a demand for them.

LARGER TYPES OF PROTEAS

These grow up to about 9 or 10 feet and are generally bushy, needing at least 8 feet in diameter in which to spread. Some only grow to a height of 6 or 7 feet, but are bushy and robust. They are suitable for backgrounds and may be used to create privacy in the garden in place of a dividing hedge. They are more suitable for large gardens than for small, but can be kept compact by constant cutting back of the branches when picking the flowers.

Proteas

P. nitida 10′
P. caffra 10′
P. compacta 10′
P. lacticolor 8′- 10′
P. laurifolia (P. marginata) 10′
P. aurea 10′
P. coronata 8′- 10′

Pincushions (*Leucospermum*)

L. catherinae 6′- 8′
L. conocarpodendrum 6′- 8′
L. reflexum 12′

Gold-tips (*Leucadendron*)

L. argenteum 15′- 20′
L. laureolum 6′
L. discolor 7′

Protea mundii 10′
P. neriifolia 10′
P. obtusifolia 8′-10′
P. punctata 8′-10′
P. repens (P. mellifera) 8′
P. roupelliae 15′
P. rubropilosa 15′

Leucadendron eucalyptifolium 6′
L. lanigerum 5′-6′
L. rubrum 6′-8′
L. coniferum 4′-10′
L. salicifolium 6′-10′

Mimetes hirtus 5′-8′
Brabejum stellatifolium 15′-25′

MEDIUM-SIZED PROTEAS

These grow from 3 to 5 feet in height, although a few will reach 7 or 8 feet if growth conditions are good.

Proteas

P. aristata 5′
P. magnifica (P. barbigera) 4′–5′
P. acuminata 3′-4′
P. cynaroides 4′-5′
P. eximia (P. latifolia) 6′-8′
P. grandiceps 4′-5′
P. lepidocarpodendron 4′-6′
P. longifolia 3′-5′
P. lorifolia (P. macrophylla) 4′-6′
P. effusa 3′-4′
P. nana 3′-3½′
P. burchellii, spreading, 3′-4′
P. scolymocephala 3′
P. speciosa 3′
P. stokoei 6′
P. susannae 6′

Leucadendron

L. salignum 1′-4′
L. comosum 4′-5′
L. daphnoides 3′
L. cinereum 4′-5′
L. tinctum 4′
L. gandogeri 4′
L. muirii 2½′
L. platyspermum 6′

Leucospermum

L. cuneiforme 3′-5′
L. bolusii 4′-5′
L. rodolentum 3′-4′
L. oleifolium 3′-4′
L. vestitum 4′
L. grandiflorum 3′-6′
L. gerrardii 1′-2′
L. lineare 4′
L. muirii 4′
L. cordifolium 4′-5′
L. tottum 4′

Serruria

S. pedunculata 2′-3′
S. barbigera 2′
S. florida 4′-5′

Mimetes

M. argenteus 4′
M. cucullatus 4′

Paranomus

S. sceptrum-gustavianum 4′-5′
P. spicatus (P. crithmifolius) 3½′
P. reflexus 4′-5′

Leucadendron xanthoconus 4´ -5´
L. pubescens 5´
L. microcephalum 4´
L. linifolium 4´
L. uliginosum 5´
L. sessile 3´

Aulax
A. umbellata 4´ - 6´
A. cancellata 6´

Diastella divaricata 18´

FOLIAGE

Proteas with Silver and Grey Foliage

Leucadendron argenteum
L. album
L. pubescens
L. uliginosum
Leucospermum rodolentum
L. reflexum

Mimetes argenteus
Paranomus spicatus
Serruria pedunculata

Proteas with Heart-shaped Foliage

P. amplexicaulis
P. cordata

Proteas with Needle-like leaves

Protea aristata
P. humiflora
P. pudens
P. nana
P. odorata
P. pityphylla
P. subulifolia
P. witzenbergiana

Aulax pinifolia
Leucospermum lineare
Leucadendron album
Paranomus spicatus
　(*P. crithmifolius*)
Serruria pedunculata
S. barbigera
S. florida

LOW OR SPREADING PROTEAS FOR ROCKERIES

Most of these grow low on the ground and spread. They are particularly suitable for rockeries and some will overhang the rocks. They can also be placed on slopes to cover a few feet of ground or used in the foreground of a shrub border.

Proteas

P. acaulos 1´ high
P. subulifolia 1½´

P. cordata 1´
P. cryophila 2´, spreading 9´

Protea amplexicaulis 2′
P. aspera 1′
P. restionifolia (tufted) 9″
P. scabra (tufted) 6″
P. sulphurea 2′,
   spreading to 15′
Leucospermum gerrardii 1′-2′
L. hypophyllocarpodendron,
   flat

P. humiflora 2′
P. pudens 2′
P. venusta 18″, spreading to 15′
P. witzenbergiana 1′-2′,
   spreading
P. lorea (tufted) 9″
Leucospermum prostratum, flat

PROTEAS FOR POTS OR LARGE TUBS

They make attractive evergreens. Drainage must be good and this makes it easy to cultivate these that need well-drained soil in areas of high rainfall, as watering can be controlled.

Protea

P. acuminata
P. amplexicaulis
   (will drape over sides)
P. aristata
P. canaliculata
P. cynaroides
P. effusa
P. nana
P. pendula
P. witzenbergiana
   (will drape over sides)

Mimetes

M. argenteus
M. cucullatus

Leucospermum

L. oleifolium
L. lineare
L. muirii
L. prostratum
   (will drape over sides)

Leucadendron

L. album
L. salignum
L. gandogeri

Serruria

S. barbigera
S. florida

HOW TO PLANT A PROTEA COLLECTION

The sketches overleaf have been made to assist the gardener in planting a collection of some of the more attractive Proteas. The choice is arbitrary and can be adapted to any collection. They have been grouped according to size, with the taller types at the back so that they do not obscure one another. At the same time, they act as an evergreen screen for privacy and can be cared for appropriately.

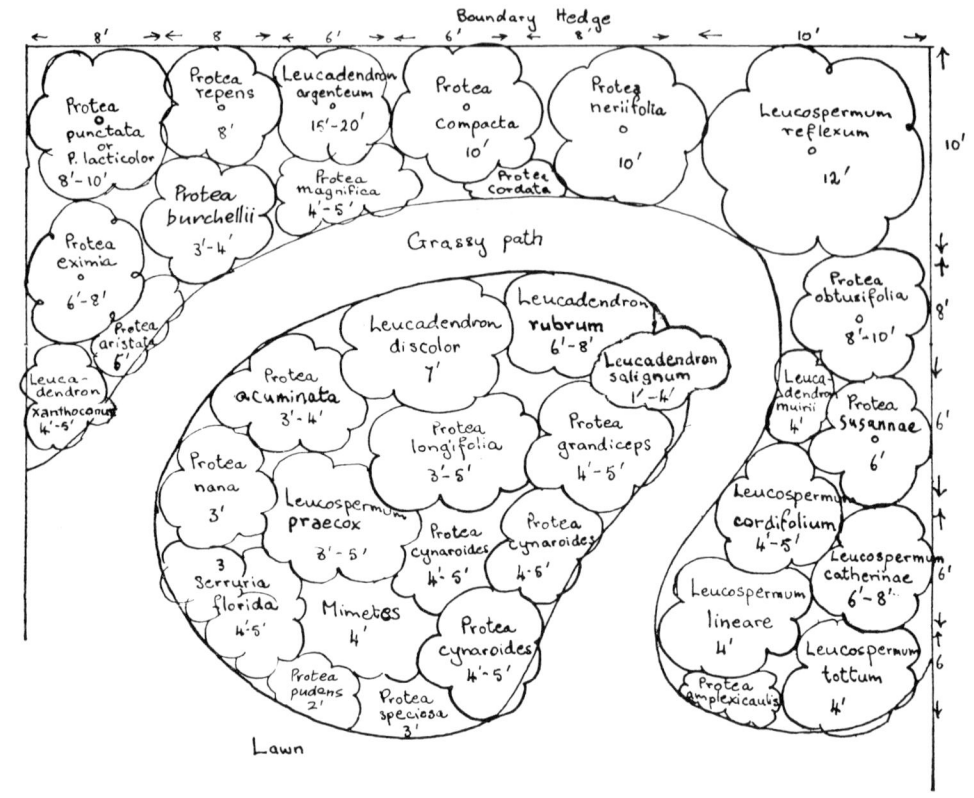

Planting Plan for a Corner Boundary—about 50′ long.

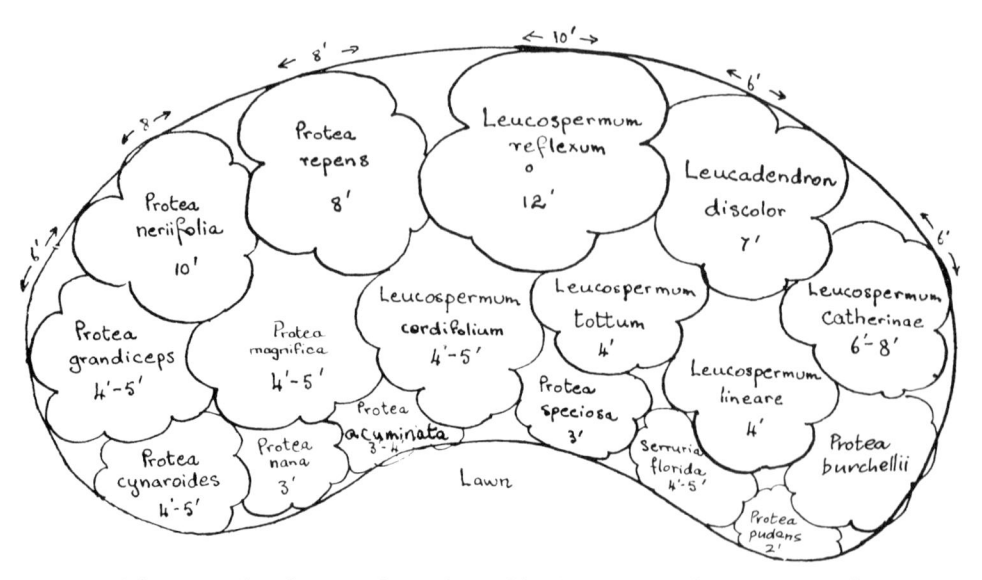

Planting Plan for a Kidney-shaped bed—about 40′ long and 18′ wide.

CHAPTER IX

PROTEAS AS CUT FLOWERS

Proteas could become one of the world's finest commercial cut flowers, along with Roses, Carnations and Gladioli, were they plentiful enough. In fact, I would go so far as to say that the Nodding Pincushion Flower (*Leucospermum cordifolium*) is the world's finest cut flower when you consider its lasting qualities—a month in the vase—its vibrant colours and its interesting possibilities for graceful and informal as well as modern and dramatic flower arrangements. It has no scent, and that may demote it to second place, but its satisfying size and its exquisite form and colour would recommend it to me in preference to the ubiquitous Carnation and stiff Gladiolus. Roses are, of course, a class in themselves and not to be compared to any other flower, for their beauty is so rich and regal that it compensates for the comparatively fleeting quality of the bloom.

What makes the Protea qualify as an outstanding cut flower? Proteas flower throughout winter and spring, particularly when most flowers are scarce, and have superb lasting qualities. They will last in the vase for at least a week, looking extremely fresh, and then continue to look decorative while drying out naturally into gentle tapestry browns and pinks that are everlasting. For Proteas never wilt and die. They simply fade into another form that can be used in interesting dried arrangements. These are extremely useful in hallways and studies where it is a relief not to have to keep replenishing the flowers. A bouquet of Proteas can remain attractive for years as a pleasant reminder of the giver.

Dried Proteas can serve a useful purpose in helping the homemaker of taste who needs some permanent flower arrangement, yet is revolted by the artificial flowers that have appeared to fill the need of those who have neither the time, the money nor the patience to arrange fresh flowers. Dried flowers are real flowers and dried plant material of many kinds evokes an appreciation of a less obvious beauty from that of a fresh flower. The curved beauty of a graceful dried leaf like that of Watsonia, or the wonderful construction of seed pods like Lotus (*Nelumbia*) and Woodrose (*Ipomoea tuberosa*), the curling grace of the ordinary Maize tassel, all these fire the imagination of the artistic person.

Some Proteas change their shapes slightly as they dry. The slim *P. speciosa* opens outwards, with each extended bract preening its furry rich brown tip, and *P. neriifolia* behaves similarly. Other Proteas seem to shrivel and grow smaller, but develop interesting textures. The outer bracts of *P. cynaroides* often take on a mushroom-coloured silkiness; *Protea magnifica* does not lose its soft blurry quality for years; the Blushing Bride becomes a papery, delicate pale pink version of the original; the green foliage of some Leucadendrons retains its colour, but becomes a soft Adam green which blends perfectly with the subtle tones of the other dried Proteas. When some Proteas are perfectly dry, the mass of hairy flowers at the centre, which are each attached to seeds, loosen and may be pulled away easily. The hard brown bracts which remain behind flatten out like stars and add a different type of decoration to a mixed vase of dried Proteas. Pincushion Flowers do not make good dried flowers as the "pins" wither and lose their identity when they die.

Cut flowers that travel well out of water are valued by florists. Fresh Proteas travel extremely well and will last for several days without ill effect. This makes them especially valuable for sending to florists and to exhibitions all over the world. A consignment of Proteas which I sent in 1964 to an international flower show in Haifa, Israel, survived a hold-up in transit so that they were out of water for four days and, in fact, some had had a previous plane journey from Cape Town to Johannesburg, so that they had been picked a week before they were exhibited, yet managed to obtain a silver medal at the exhibition, where they were in competition with exotic orchids from London and Singapore and the magnificent spring flowers of Holland. In 1958 I sent a box of flame-coloured Pincushion Flowers *(Leucospermum cordifolium)* from my own garden, together with two friends, the late Dr. G. W. Reynolds and Mrs. M. Canin, who sent Heaths and Lilies respectively, to the New York International Flower Show. Our joint entry received a Gold Medal. In the following year, an entry from the National Botanic Gardens at Kirstenbosch won another Gold Medal at the New York show, to prove once more that Proteas had been admired and appreciated. In 1960, Blushing Brides *(Serruria florida)* created tremendous interest at the Chelsea Flower Show in London. Proteas were part of the prizewinning South African exhibit at the Paris International Flower Show in 1964 and there are many other stories of success and friendship for the unique Proteas of South Africa.

Some Proteas last better than others. Those which travel extremely well include *P. burchellii, P. obtusifolia, P. speciosa, P. cynaroides, P. magnifica, P. coronata* and many others. The foliage of some goes off quickly. *Protea compacta*, for example, has leaves that turn brown within a few days of picking and spoil the effect of the blooms. Proteas which open wide quickly like *P. lacticolor* and *P. aurea* also have a limited life as cut flowers, for they lose their original beauty which depends on fresh, glowing colour and the delicacy of the hairlike styles. They are delightful when fresh, however, and the clear pink *P. lacticolor* is most attractive in a low bowl when it is picked with short stems. Some Proteas have an untidy appearance from the beginning such as *P. lanceolata* and *P. welwitschii,* so that they cannot even be considered attractive in the vase, and their lasting qualities cannot be taken into account.

There are various methods of making cut flowers last longer. In the case of flowers with woody stems, one usually bangs the ends lightly with a hammer so that the stem is slightly crushed in order to accelerate the intake of water in the vase. This is seldom done with Proteas as they usually last so well, but it does assist to crush the stems, especially if they have been out of water for a long period before arranging. Chemical preparations like "Everbloom", which are added in solution to the water in the vase, may be valuable for Proteas. It may enable one to pick Proteas in tight bud and yet ensure their opening in the vase.

As with other flowers such as Roses, Proteas will not open if they are picked too early and are too tightly closed. If they are picked in bud when they show signs of opening naturally, then they will continue to open out in the vase. *P. lacticolor* and *P. aurea* are always picked in bud and will open in the vase.

Packing Proteas for travelling has been a subject of experiment and it has been found that Proteas travel best if each bloom is wrapped loosely in tissue paper and placed dry in a box. Dampness can cause mildew and polythene keeps in the moisture. The main purpose of wrapping the blooms in paper is to prevent them from rubbing against one another and one can achieve this also by placing wads of crushed tissue paper between the blooms. The flowers should be packed closely enough to prevent movement in the box.

Cut flowers are valuable when they provide not only colour

and decoration, but interest and originality in addition. Proteas seem to have all these qualifications.

There are extremely colourful Proteas in all shades and tones. Deep reds occur in colour forms of *Protea magnifica, P. cynaroides, P. repens* and others, and wine reds are present in *P. nana, P. effusa* and *P. acuminata*. There are many hues of salmon and pink as well as clear whites, creamy yellow and shades of green amongst Proteas. Pincushion Flowers (*Leucospermum*) are even more vivid with their bright tangerines, oranges and yellows, some verging on scarlet and others scarlet-tipped.

Many Proteas are decorative as they are a good size and one does not need a great number to make a show. Arranged in masses, they make stunning decorative pieces, as anyone who visited the wild flower show at Goodwood, Cape Town, in celebration of the Jubilee of Kirstenbosch in 1963, will remember. On the other hand, a single beautiful Protea in a specimen vase will be a conversation piece in itself. Even the smaller types make dainty subjects in a slender vase.

From the aspect of interest and originality, Proteas are winners. The different types are so numerous that they create interest in many ways. Rarer types which are not seen often are fascinating to those who endeavour to recognise their names as well as to those who are ever seeking the unusual. People who delight in arranging flowers find a challenge in arranging uncommon types like *P. acaulos*. Florists find pleasure in plants with interesting and useful foliage like many Leucadendrons. Interest is always created by the unusual and it is fun to recognise the rarer Serurrias or other members of the Protea family and discover their places of origin.

Proteas can make very original flower arrangements for they may be used in many different ways. The conventional way of arranging Proteas is to group them in a free, full or informal bouquet. They can also be used in formal arrangements, such as the crescent-shaped curve in the picture on the frontispiece or with dramatic effect.

Copper makes an excellent foil for Proteas and they are extremely easy to arrange in a copper jug or deep vase. Containers with character like kettles, brass samovars and antique iron pots are most effectively filled with the rich yet mellow colours of Proteas and the Protea family. They blend harmoniously with richly glazed simple pottery in soft blues, greens, cream and brown, but do not look quite so much at home in pure white or black modern vases or in brightly coloured containers.

Proteas are rugged in character and lend themselves to interesting displays that might not suit daintier flowers. Yet one cannot generalise and one can use the daintier types like *Protea nana* and the Blushing Bride (*Serruria florida*) in dainty glass or in a dull white classical urn.

Do not refrain from mixing Proteas with other flowers for fear that they might seem incongruous. I have seen Proteas mixed successfully with coral or pale pink Carnations and one of the most attractive mixed bowls I ever had contained a combination of rosy *Protea burchellii*, stripped of foliage and cut short, Iceland Poppies, yellow roses, tawny autumn foliage of *Hypericum* and short sprays of orange *Bougainvillea* MacLean, stripped of foliage.

Proteas are often half hidden by enclosing leaves around the flower-head. One should clip away some of the leaves to expose the colours, but try to keep enough foliage to prevent the look of a hen without its neck feathers. If the foliage has gone brown or if one needs to mix the Proteas with a mass of other flowers, the leaves may be removed entirely and the naked stems "masked" with the other plant material.

Dramatic modern arrangements can be made, using shallow bowls and heavy pinholders. I once saw a round copper tray on a low coffee table arranged with an asymmetrical wheel of reddish Pincushion Flowers (*Leucospermum cordifolium*), radiating from a point off-centre to beyond the rim encircling the tray. The point of emergence was camouflaged with two or three short-stemmed blooms.

Again using a shallow container, one could place a short-stemmed large Protea or Pincushion on a pin-holder and back it with upright, uneven lengths of yellow bamboo, the stiff leaves of New Zealand Flax (*Phormium tenax*) or the large leaf of a Cut-leaf *Monstera*. An oblong shallow container with an upright spray of three or five Proteas or Pincushions may be balanced on the other side with a large decorative stone such as a blue-green copper stone, a jagged lump of rock crystal or a large chunk of white coral.

It seems obvious, therefore, that the possibilities of arranging flowers using Proteas are endless if one is lucky enough to have them growing in one's own garden or can obtain them easily from growers or florists in one's own city.

ON PICKING PROTEAS

Do not refrain from picking Proteas freely. Picking is good for the plant in two ways. It shapes the bush by cutting down lengthy growth and keeping the whole bush compact. It also prevents untidiness by removing the dead flowers, for Proteas do not fall off by themselves and the dead brown heads look unsightly as the seasons pass. Even if the blooms are not picked with a length of stem, the dead heads should be removed from the bush by twisting them off.

Proteas often nestle at the base of three or four long branches and the gardener often hesitates to pick the whole candelabra of branches in order to obtain one bloom. This does the bush no harm, however, and the flower should be picked with a length of stem, from which point new branches will spring. The branches around the flower-head may be clipped away neatly, in order to expose the flower-head, together with some of the leaves. This habit of growth is particularly common in *Protea neriifolia* and *Protea burchellii*.

Pincushion Flowers do not remain on the bush, but wither and disappear after the flowers are spent. This means that the bush is seldom unsightly, but the flowers can also be picked with long stems without harm to the plant. In the case of *Leucospermum reflexum*, the Rocket Pincushion, which sends very long growths up to a height of about 9 feet, it is an advantage to cut the topmost flowering stems right back in order to keep the bush compact.

Pincushion Flowers often form double and triple buds which produce twin flower-heads. If a single bloom is preferred, one can disbud them while the buds are tiny. This is not always advisable in areas with cold winters, however, as the first bud is often killed by the frost, while the later incipient buds open into beautiful flowers in the spring.

PART II

KING PROTEA, GIANT PROTEA, REUSE SUIKERBOS
Protea cynaroides

Possibly the most spectacular flower in the world, this varies greatly in size according to the locality in which it is found growing wild in nature. The largest grow up to fully eleven or twelve inches across, exciting the wonder of all who see them, while some are half the size, and there is also a miniature form. It was chosen as South Africa's national floral emblem in 1976.

The wide-spaced pointed bracts of the King Protea are ringed evenly around the central snowy peak of hairy flowers. These vary in colour from a soft silvery pink to a deep clear pink and, sometimes, a deep crimson, which is not so common. They bloom from midsummer to autumn and spring. The leaves have a distinct stalk. In some cases the leaves are oval and sometimes pointed, measuring up to 2 or 3 inches across. A journalist who saw the King Protea at an International Flower Show overseas described it as "half-way between a daisy and an artichoke", and this is interesting as it was named after "*Cynara*" the Artichoke. Such a mundane description, however, does little justice to this most majestic species which deserves a place in every garden.

Protea cynaroides forms a comparatively small bush for such a giant flower, for it grows from 3 to 5 feet in height and can fit into the smallest garden. It could be grown in a large tub on a sunny verandah and put under shelter for the winter if the climate is very cold. It will survive temperatures which fall to zero, however, if it is given a protected position. Even if the plant dies back as a result of frost, new shoots often grow up from the base in the spring. Regular watering throughout the year is best for *P. cynaroides* and it may be transplanted more easily than most Proteas, even at a height of about 2 feet. It does better in good garden soil than in poor soil and does not seem to require such sharp drainage as most Proteas.

The natural haunt of the King Protea spreads over a wide area of the south-western Cape, from Clanwilliam to Grahamstown, and it is common in the Knysna area where there is regular rainfall throughout the year. Pale, silky pink forms come from the Breede River area, while deep red ones, which are among the best, come from the Outeniqua range.

P. cynaroides is extremely adaptable in gardens and grows equally well inland as on the coast, thriving even on the sub-tropical South Coast of Natal, 5 miles from the sea near Port Shepstone. They have bloomed after 18 months in this warm area, but the bushes flower themselves to death and do not last for many years.

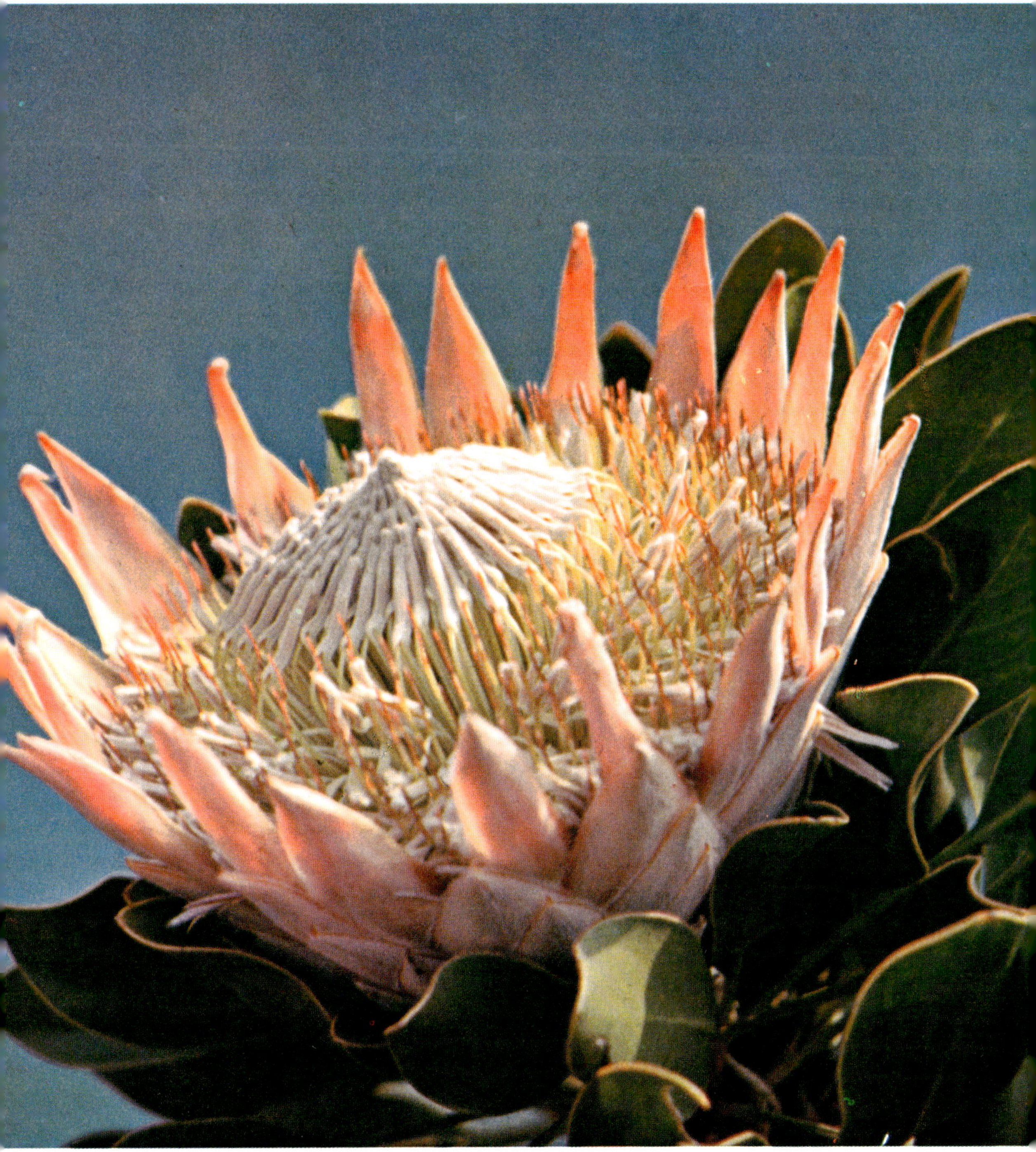

GIANT WOOLLY-BEARD, QUEEN PROTEA
Protea magnifica (*P. barbigera*)

The appeal of this most exquisite flower must be seen in reality, for no photograph can impart the softness of its furry centre or the subtle yet pellucid colouring of its bracts. Magnificent in size, for the flower-head measures from 6 to 8 inches across, it is also alluring in its colouring. The most commonly seen colour is a soft pink, which varies on different plants and sometimes becomes a deep rose or strong red. A pale yellow or deep cream colour variety is also most attractive. The centre of the flower-head is filled with a mass of white hairs which taper into a black velvety centre and each scale is edged with fluffy white hairs, The total effect is that of a glamorous powderpuff.

The Giant Woolly-beard is most effective in floral displays and always excites admiration from everyone, even if they are not Protea lovers. The blooms open slowly from midwinter until early summer, from June to November. The bush itself grows only to about 4 or 5 feet in height and has a spreading form. It is particularly effective on sloping soil as it hugs the ground, covering it with soft grey-green foliage. The leaves are about 6 inches long and over an inch wide, edged with hairs.

This species grows slowly and is not easy to grow in the early stages, for it damps off easily. It usually flowers in the third or fourth year from seed. *P. magnifica* is fairly cold-resistant, especially the yellow-flowered form that is found near Ladismith, Cape, where it is near the snowline at altitudes of about 5,000 feet. The fleecy buds remain closed for a long period in winter and the flowers are at their very best in spring. They grow wild on the Hottentots Holland Mountains and other mountains of the south-western Cape near Tulbagh and the Hex River Valley, Worcester, Ceres and Clanwilliam. They grow in rocky places in nature so that drainage is of great importance in their cultivation.

The older name, *magnifica*, must now replace *barbigera*, as published in 1976 by Dr. J.P. Rourke, who is revising the genus. (See *The Journal of South African Botany*, Vol 42. Part 2 of 1976)

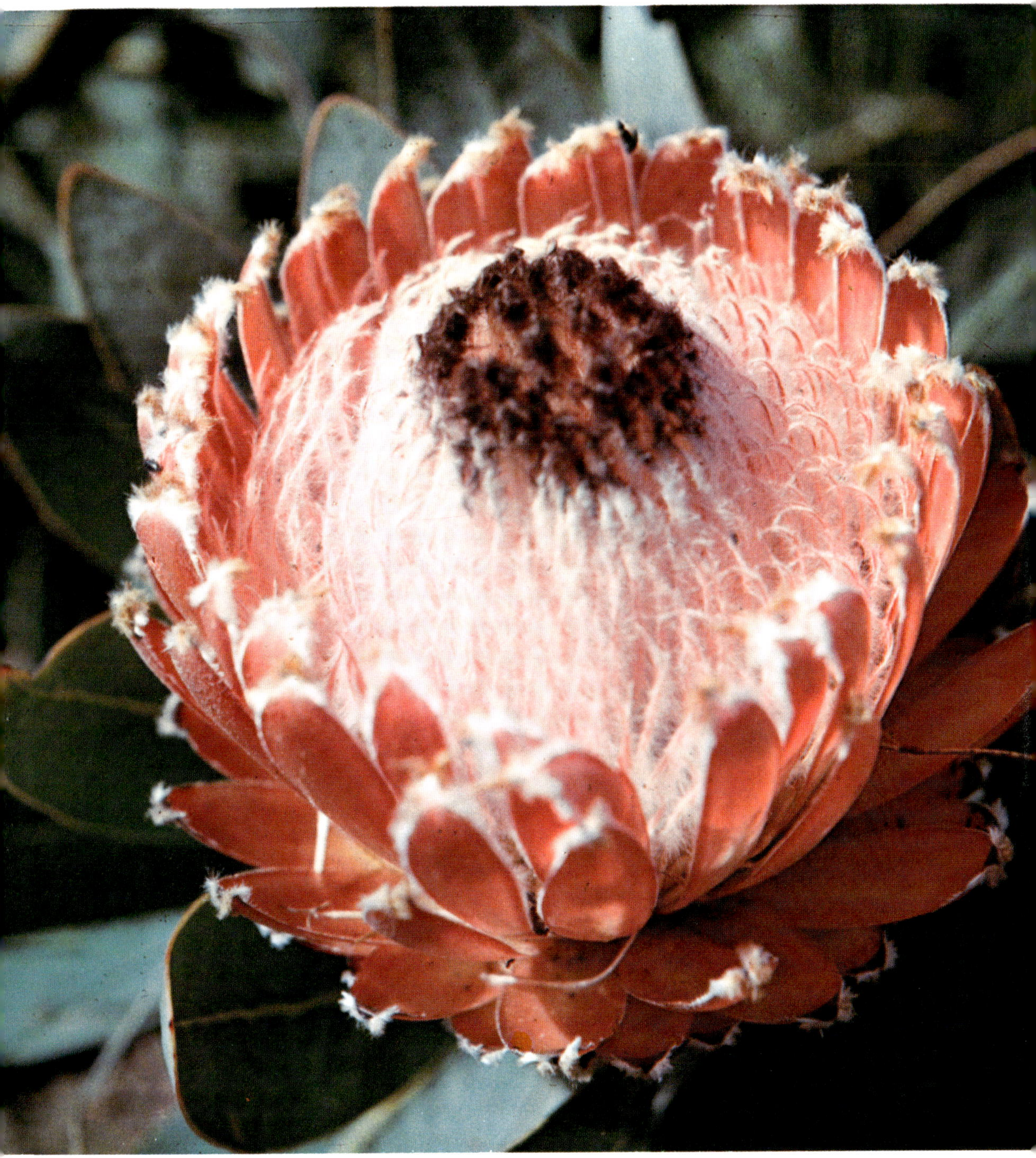

TRUE SUGARBUSH, WARE SUIKERBOS
Protea repens (*P. mellifera*)

This Protea is the famous Sugarbush that was so well-known to the early colonists, who used its nectar as a source of sweetening, from which they made Suikerbos Stroop, and which has been immortalised in a traditional folksong. It is easily recognisable, for its long pointed bud glistens with sticky nectar, which heightens the precise effect of its stiff overlapping scales. The well-defined and pointed bracts are smooth and have no hairs. They are shaded with rosy tips which vary in tone from pale pink to strong red. There is also a pure white form of *P. repens*, which is particularly lovely when fresh and unblemished.

The bud is about 5 inches long and only about $2\frac{1}{2}$ inches in diameter, but it opens fairly quickly into a V-shape. *Protea repens* starts blooming early in autumn and the flowers appear at intervals throughout winter until spring. The foliage is distinctive, for the leaves are long and narrow, giving a light airy appearance to the bush. The shrub itself is large and rounded, growing up to about 9 feet in height. It needs at least 6 feet in which to spread. Dwarf types sometimes appear in nature and there is a 4-foot form with pink or white flowers that occurs near Ladismith, Cape. It grows near the snowline, indicating that *P. repens* is one of the more cold-resistant types.

When he first named it in 1753, Linnaeus gave it the name of *Leucadendron repens*, meaning "creeping" in Latin. He had based his description on the drawing that appeared in Boorhaave's work of 1720, never having seen the plant in nature or even a real specimen. He later changed this name to *Protea repens*. Thunberg, during his stay at the Cape from 1772-1775, saw the Sugarbush in nature and called it, appropriately, the honey-bearing Protea, which is *Protea mellifera* in Latin. Thus it was known until 1960, when Prof. H. B. Rycroft discovered the earlier name. According to the International Code of Botanical Nomenclature, the first specific name applied after 1753 must be retained, so that this charming name had to be dropped in favour of *P. repens*. Gardeners do not take kindly to changes like these, but the rules must be followed so as to avoid confusion. At any rate, the common name of Sugarbush will always remain to popularize this species.

In the early days, *P. repens* grew plentifully on the lower slopes of the mountains of the south-western Cape, but it is becoming rare through the spread of civilisation. In Mrs. Roupell's work of 1849 it was called "the ever-cheerful sugarbush" and referred to as the most common species of Protea, in bloom for 8 or 9 months of the year. As it grows easily in cultivation and is such a popular cut flower, one hopes that it will never become extinct.

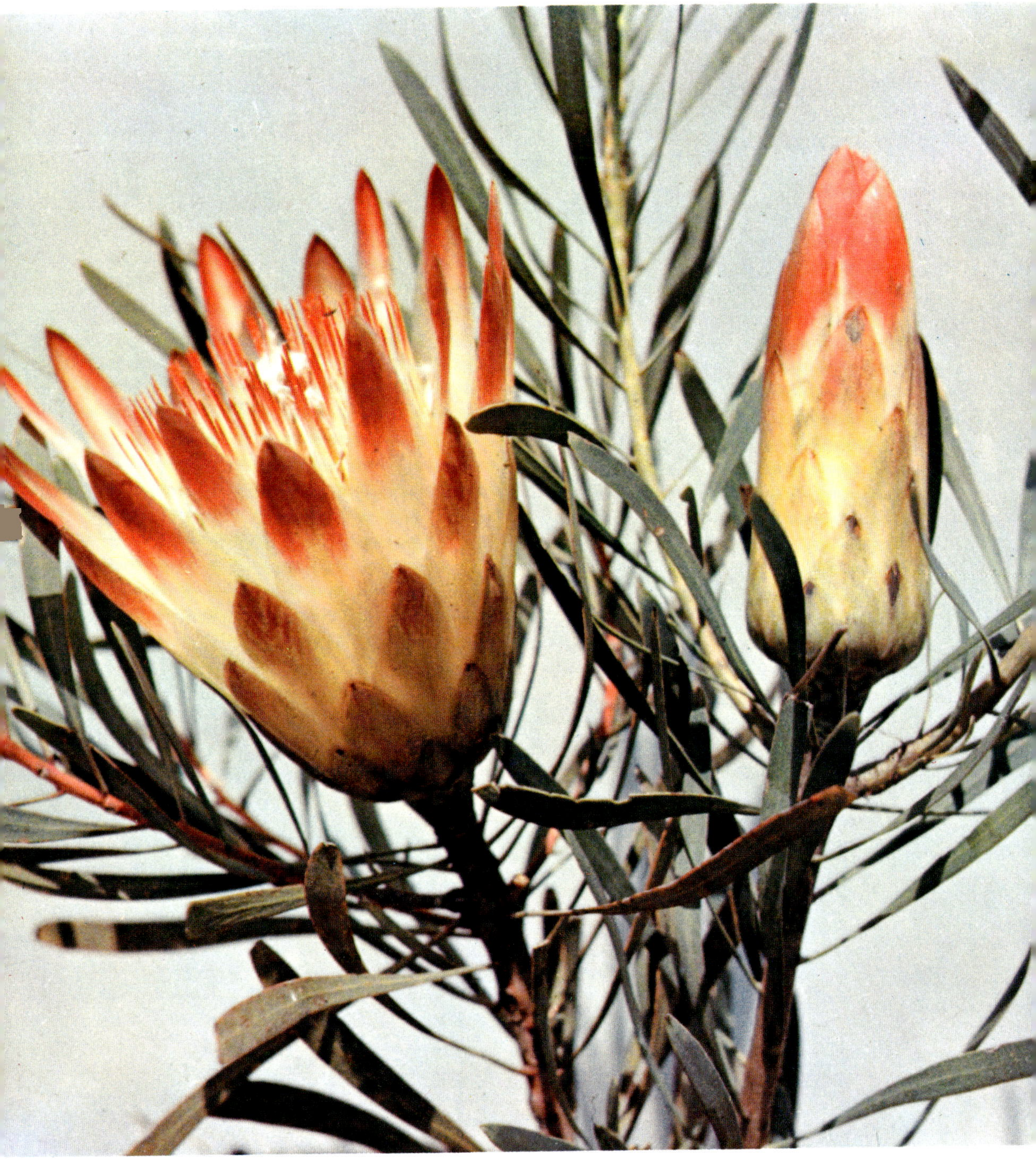

BREDASDORP SUGARBUSH
Protea obtusifolia

Similar in many ways to *P. repens*, this species has a more intricate, shorter flower-head and broader leaves than the true Sugarbush. The overlapping, smooth bracts are fairly closely set and they have a distinct greenish hue, each one strongly tinged with colour, which may be pink or a deep pomegranate red. A white form is also available, as may be seen in the illustration. The flowers appear during winter and spring and become very shiny with nectar.

P. obtusifolia forms a large, rounded woody shrub which reaches a height of about 10 feet when conditions are favourable. The green leaves can be up to an inch in width and about 4 inches long.

Apart from the fact that *P. obtusifolia* grows very easily in the garden, often starting to flower when only 18 months old and less than 2 feet in height, it is remarkable and interesting because it grows in limestone soil in nature. It is plentiful in the Bredasdorp outcrop, from Stanford to the Breede River mouth. It grows rather poorly in stony ground, with much smaller flowers, so that it responds to good soil in the garden. *P. obtusifolia* is very tolerant and will grow in ordinary garden soil even of an acid or clay nature. It should be among the first species to try in countries which have alkaline soil, where most other Proteas fail.

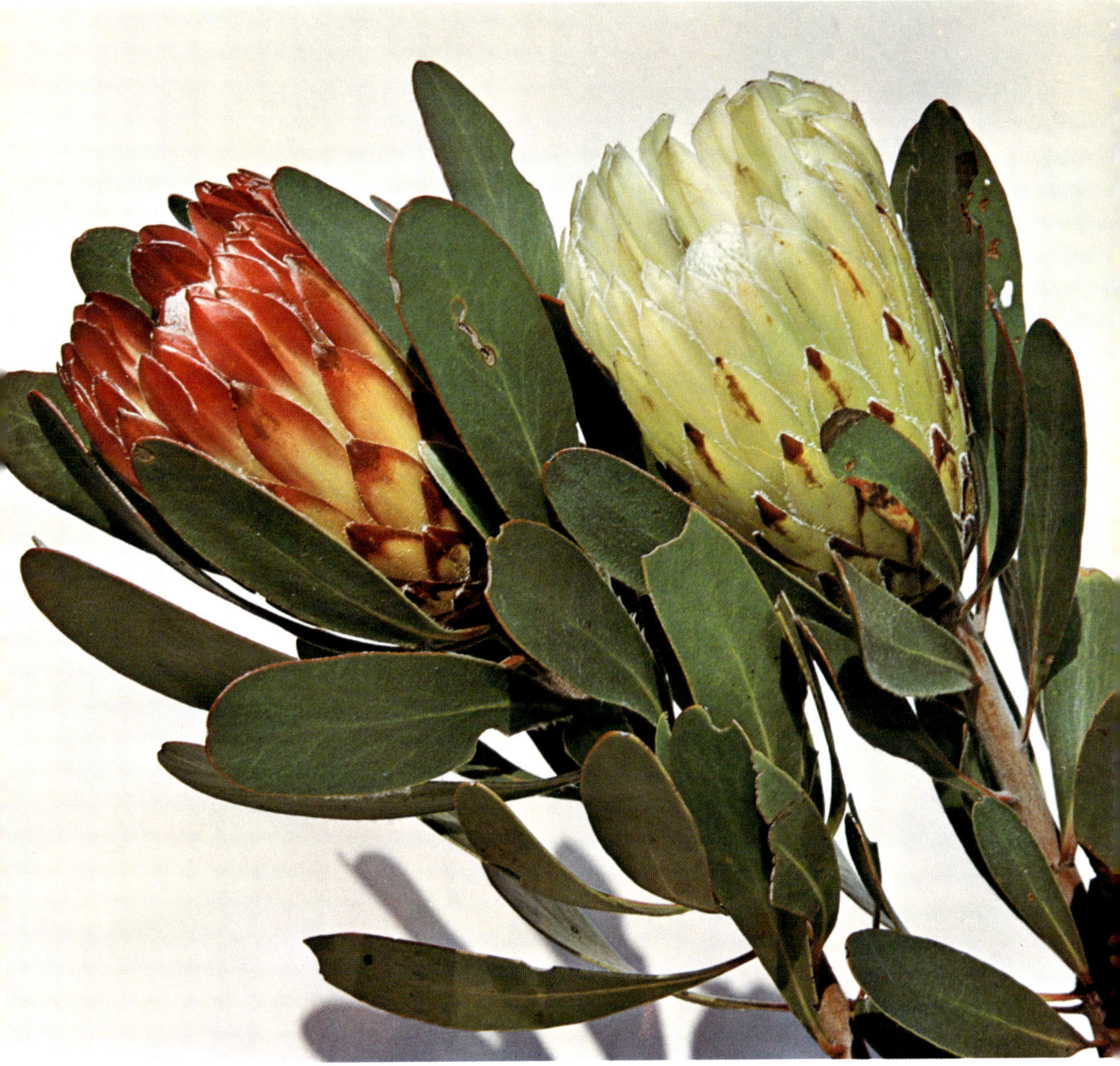

WABOOM

Protea nitida (P. arborea, P. grandiflora)

In the neighbourhood of Cape Town, and in the mountains of the south-western Cape, the tree-like Waboom was common in the early days of Cape history. The settlers used it freely for firewood and the bark was used for tanning. Its common name of "Wageboom" or "Waboom" was derived from the fact that the colonists used its wood to make their wagon-wheels. Ink made from its leaves was said to have been used by Louis Trichardt when he wrote his diary. Its popularity as a useful economic tree made it less common as the years went by and *P. nitida* is only now being respected and protected in nature.

The tree grows up to about 10 feet in height and can even reach 20 feet in favourable conditions, as it does on Table Mountain. It has a gnarled and stunted appearance when it grows in rocky places without depth of soil.

The flowers appear during autumn and winter. The heads are large, up to 4 or 5 inches across, and while some may think them colourless, they have a beautiful silken finish and their yellow-green colour has an air of elegance. The centre is filled with stiff creamy hairs. The smooth, hard foliage is greyish-green and offsets the delicate colouring of the blooms. These are striking flowers in the vase, although they do not have the conventional beauty of the more showy Proteas. There is a variety with very broad leaves, and also a rare variety with red flowers.

Despite its hardy appearance, *P. nitida* should be given the same cultural conditions as other Proteas from the winter-rainfall area. It is not likely to become a popular horticultural species, but should be given a place in a large garden or in a collector's garden.

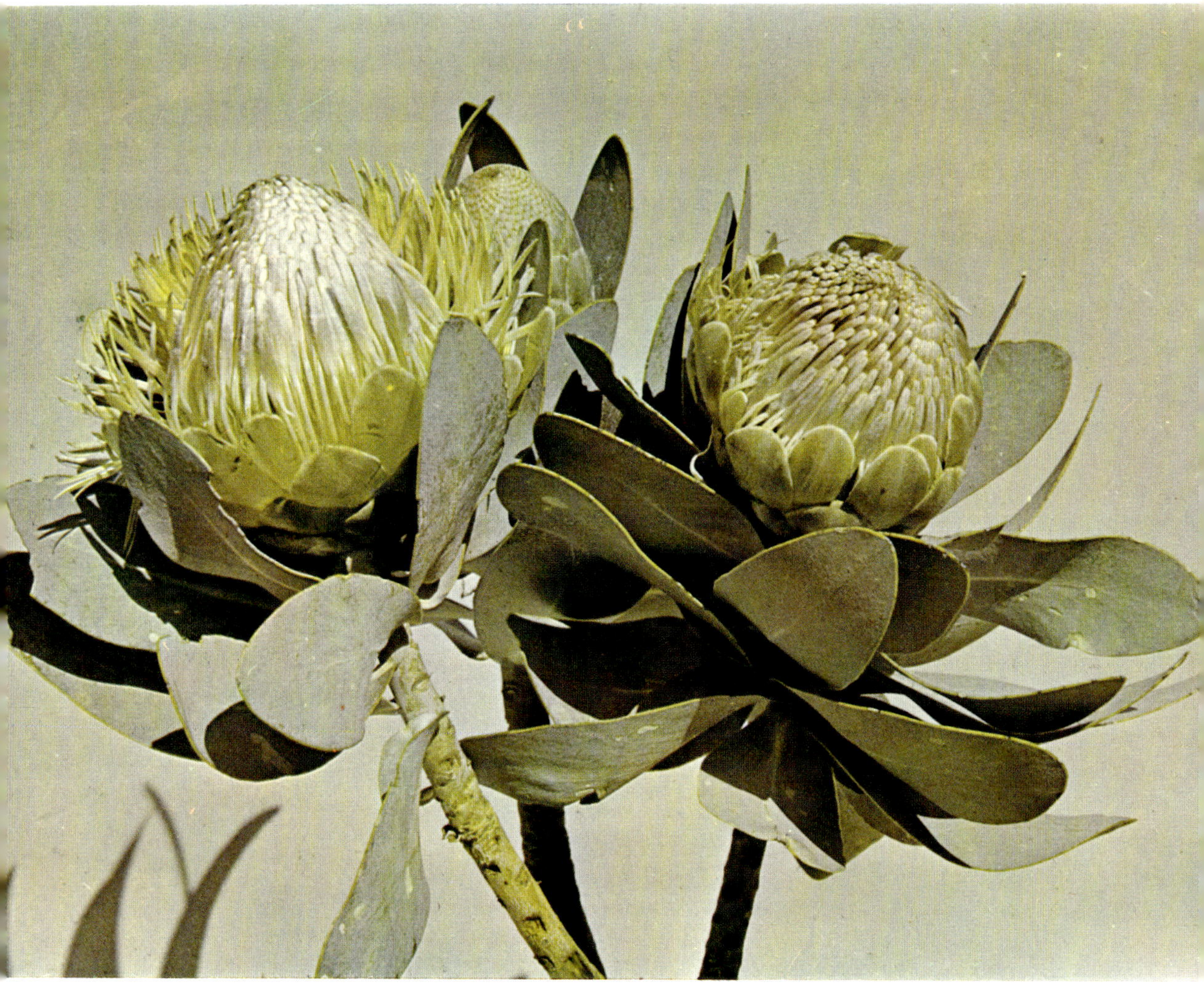

BROWN-BEARDED PROTEA
Protea speciosa

The delicate colouring and satiny texture of this beautiful flower make it a favourite of florists and gardeners. It is easily recognised by the thick, brown silky beards at the tip of each coloured bract. These bracts may be bright pink, an exquisite shell-pink or, more rarely, creamy yellow. Their silvery sheen is due to the fact that they are covered with short, pale hairs. The bud is long and fairly slim, about 5 or 6 inches in length and 3 inches across. The combination of tawny brown and pink makes this Protea distinctive and enchanting. The flowers appear in late summer, autumn and, sometimes, in winter. They bloom in December in Betty's Bay.

The name "Prince" has been suggested for it by an exporter of cut flowers in an effort to popularise it abroad. It is popular with florists in this country, but the foliage is generally stripped as it is frequently nibbled and spoiled by insects. The leaves are broad and green, about $4\frac{1}{2}$ inches long and sometimes up to 2 inches wide. They have a leathery texture. The bush itself is sturdy and grows to about 3 feet in height.

This species is well worth cultivating and is obtainable in the trade. Germination from seed is good. It is fairly cold-resistant and needs regular watering during winter. *P. speciosa* grows wild in the mountains of the south-western Cape and also grows near the sea at Gansbaai. It should be compared to *P. stokoei* and *P. patersonii*, which have similar flowers. An unusual form of *P. speciosa* that grows on Muizenberg Mountain has thick white fur on all but the topmost bracts, which are brown-tipped.

A form of *P. speciosa*, formerly known as *P. patersonii*, has a similar pink flower to that of *P. speciosa*, with densely, woolly-tipped bracts, but the head is a little smaller. It can be recognised instantly by its long slender leaves that are 3 to 5 inches long and only half an inch wide. There are no veins on the under surface. This spring-flowering Protea grows wild only in the Caledon area of the south-western Cape.

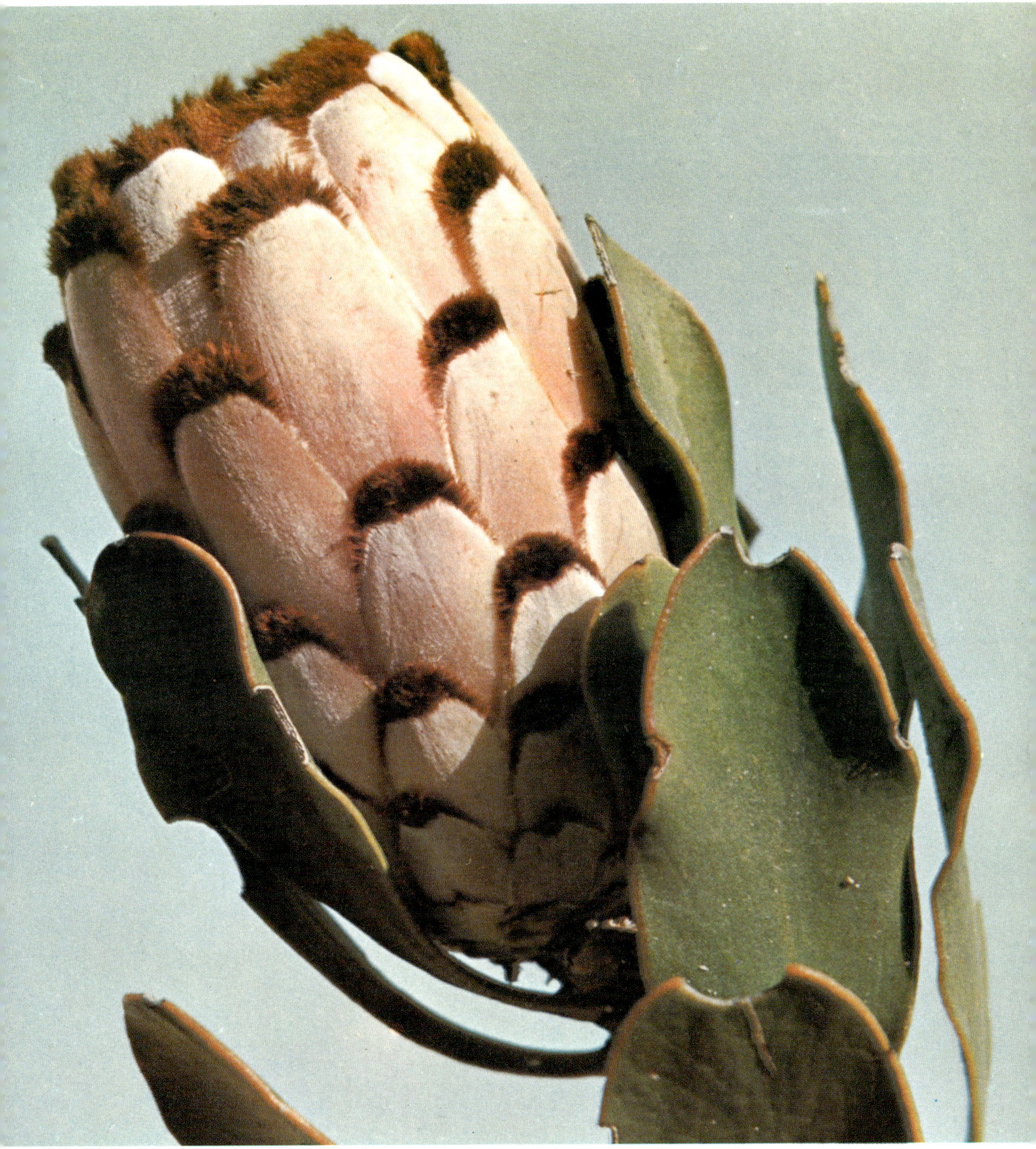

PEACH PROTEA, OVAL-LEAVED PROTEA
Protea grandiceps

One of the most glorious flowers in the Protea family, this is seen at its best when the arched bracts are not quite fully open and before the central white hairs are exposed. Each coloured bract, varying from a peach or salmon-pink to a deep coral-red, is edged with a fringe of long, silky, reddish-brown and white hairs that rest like eye-lashes on the mass of hairy white flowers inside. The flower-head measures almost 4 inches across and does not open wide. It is about 6 inches in length. A popular name, "Princess", has been suggested for this species.

The flowers are at their best in late spring and continue in bloom until midsummer. They appear in profusion at the tips of the upright branches, each encircled with a whorl of oval, greyish-green leaves. These are distinctively edged with a red line and measure nearly 3 inches in width and about 4 or 5 inches in length. The shrub grows neatly in a rounded shape of about 4 or 5 feet in height.

P. grandiceps grows slowly, but is so beautiful that it should be grown frequently and given a favourable position. It usually flowers in the fourth year from seed and the blooms appear on the small plant, which does not reach its full height for about eight years. It needs the usual care given to Proteas from the winter-rainfall area and is quite cold-resistant. It grows wild in high places on the mountains of Du Toit's Kloof and in the Worcester and Swellendam Divisions. It was orginally discovered by Niven on Devil's Peak, but has been almost exterminated there by fires and by the spread of the Cluster Pine (*Pinus pinaster*). *P. grandiceps* is rare in nature, where it should be protected, but plants are obtainable in the trade. The day may come when it is grown so plentifully that plants can be transferred to its native haunts.

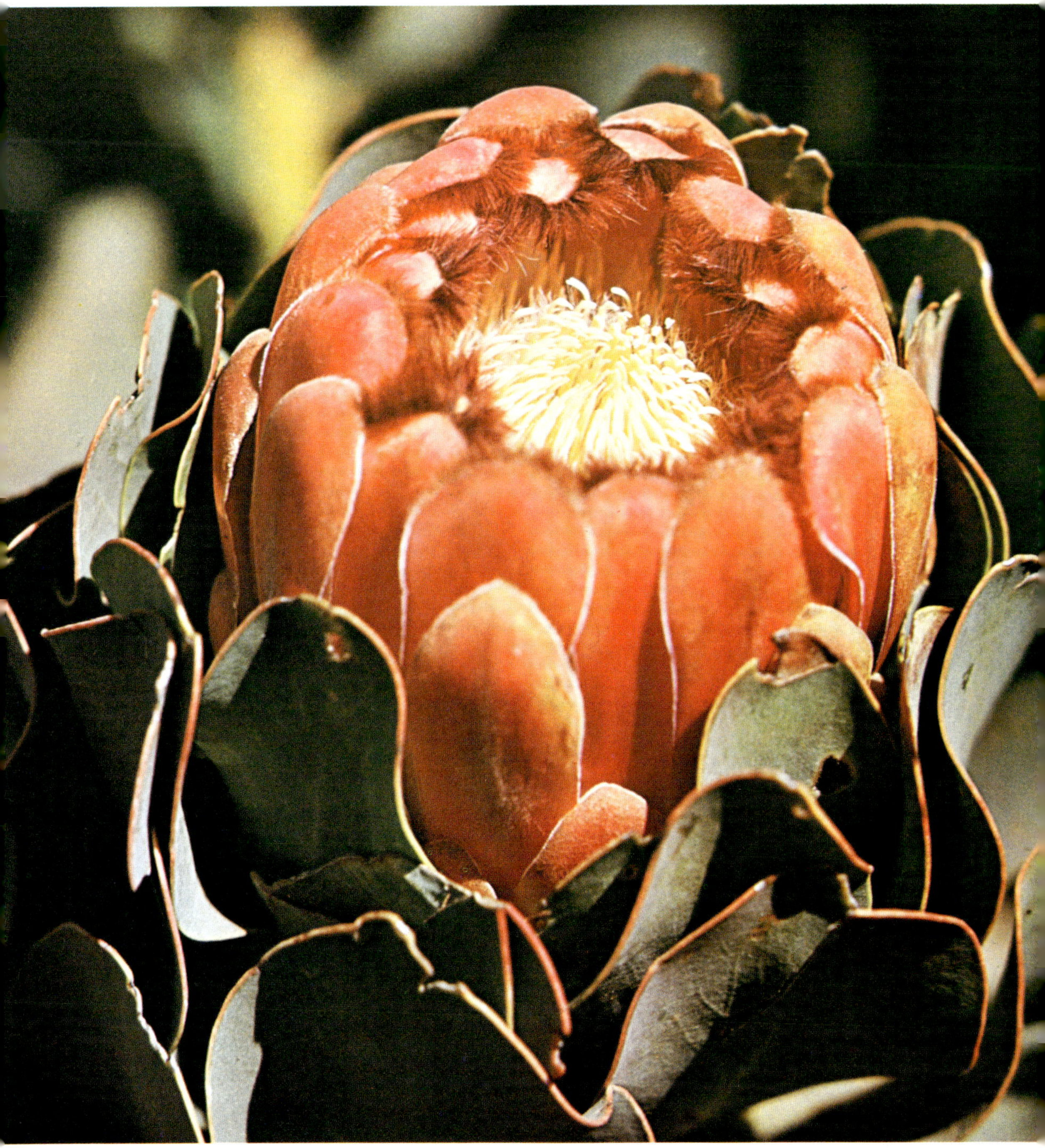

OLEANDER-LEAVED PROTEA, BLOUSUIKERBOS
Protea neriifolia

One of the best Proteas with which to start a collection, this is quick and easy to grow, flowers profusely and makes an excellent evergreen shrub in the landscape.

The flowers, which are produced in hundreds during autumn and winter, have a striking dark beauty. Each head measures about five inches long and three inches in diameter, seldom opening wider. The topmost satiny bracts are tipped with a beard of black or purplish fur. There are several colour varieties, ranging from deep pink to pale salmon, but all have the same furry tips and silky sheen. The lighter colours are sometimes more effective than the dark. The central mass of hairy flowers is set deep in the cup and generally coloured pink, tipped with reddish-purple, but it may be creamy-yellow. The colour of the hairs on the bracts and beards has been used to distinguish this species from the very similar *P. laurifolia,* but this is not a reliable feature and they are best differentiated by their leaves.

The leaves of *P. neriifolia* have been likened in shape to those of the oleander *(Nerium oleander)* as they are long and narrowly oblong. They seem to be cut off squarely at the base (truncate) and are generally attached to the stems without leaf-stalks or petioles. They are green in colour, velvety in texture and parallel-veined. The leaves of *P. laurifolia* are smooth, greyish-green in colour and distinctly broader in the middle, tapering to the tip and at the base to form a distinct leaf-stalk (petiole) that is 5-10 mm long.

The two plants have a different appearance in the field. *P. neriifolia* forms a large shrub of 3 metres in height and 2 metres across, that branches a short distance above ground level, while *P. laurifolia* is a small tree with a fairly prominent, though short, trunk.

P. neriifolia is hardy to about 15 degrees of frost, but will be killed by very severe weather. The flowers are often browned by frost during winter, even though the bush remains unharmed. In areas with cold winters, therefore, it should be given shelter from the morning sun. It grows in a wide variety of soils, preferring acidity and good drainage, but will tolerate lime and clay soil. It needs fresh air, sunshine and regular watering in winter. By the time the flower-head is ready for picking, it nestles amongst new growth that has sprung up from the sides, but one should not hesitate to cut it all off together with the flower, as the bush replenishes itself easily.

This species grows wild in the south-western Cape, on the slopes of the mountains near Paarl, Stellenbosch, Caledon, Bredasdorp, near the Breede River, and eastwards to Port Elizabeth.

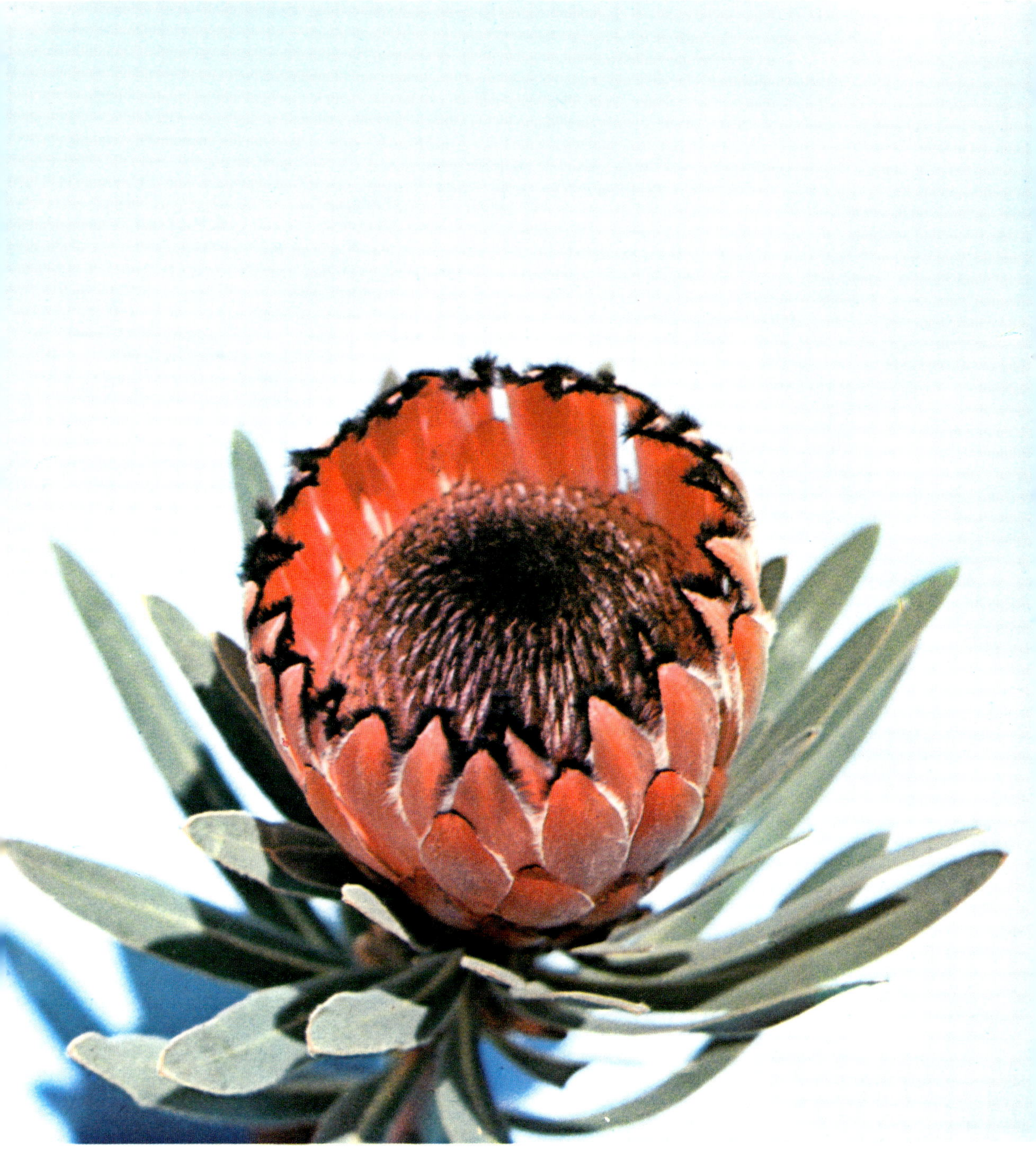

LAUREL-LEAVED PROTEA, FRINGED PROTEA
Protea laurifolia (*P. marginata*)

It is difficult to distinguish between *P. laurifolia* and the better-known *P. neriifolia,* but it may be recognised by its broader, greyish-green leaves, which are similar in shape and size to those of the laurel, tapering at both ends and with a distinct leaf-stalk or petiole, of 5-10 mm in length. The beard on the tips of the bracts generally has an intermingling of black and white hairs and the flower-head itself is similar to that of *P. neriifolia*. The bracts are a silvery pale pink that is very distinctive, and there is not much variation in colour as there is with the Oleander-Leaved Protea. The central mass of hairy flowers is a delicate yellow, while that of *P. neriifolia* is usually purplish.

The Laurel-Leaved Protea blooms mainly in the spring and is certainly one of the attractive species to plant in the garden. As it becomes a very large, rounded bush or small tree, growing up to about 10 feet or more, it should be given at least 8 feet in which to spread. It has a prominent, though short, trunk, thus differing from the bushier *P. neriifolia*. It could be extremely useful as an evergreen screen on the boundary or form part of a background to the herbaceous border. *P. laurifolia* grows easily in ordinary, well-drained soil. It is fairly drought-resistant and stands up quite well to cold weather. It grows wild in the south-western Cape, being found near Clanwilliam, Bains Kloof, Tulbagh, Paarl and Swellendam, as well as the Port Elizabeth area.

P. lorifolia (*P. macrophylla*) is apt to be confused with *P. laurifolia* as it is pronounced in the same way. This is a bush of 4 to 6 feet with a flower which is about the size of *P. obtusifolia*. The colours come in a wide range of shades of pink, biscuit and brown. The distinctive leaves are very long, growing up to eleven inches in length, but are only about an inch in width. *P. lorifolia* grows wild in the Swartberg mountains of the Cape.

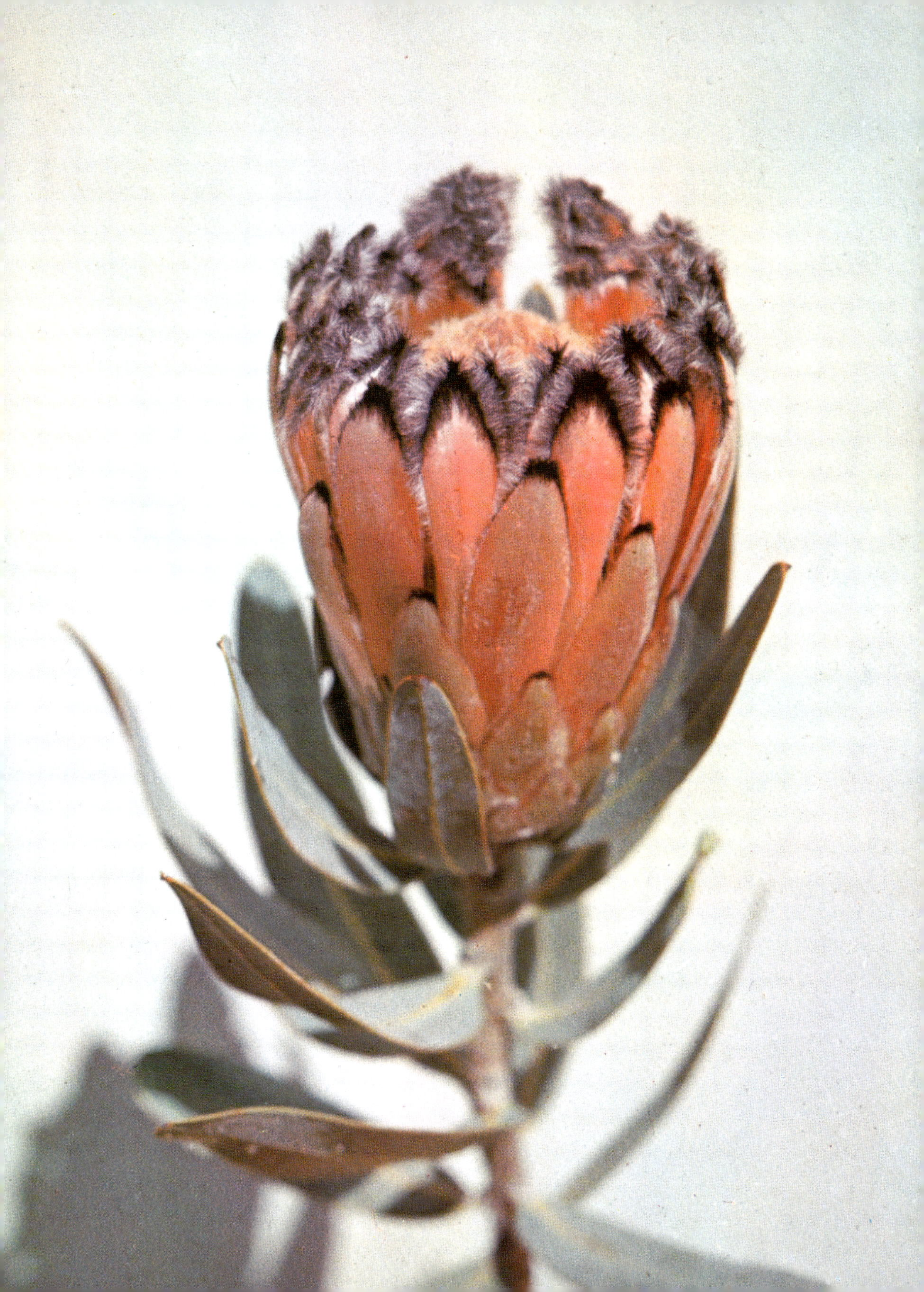

HIGHVELD PROTEA

Protea caffra

A well-known tree on the Witwatersrand, this is seldom cultivated in gardens, yet it is always retained by home-owners who find it growing wild on their property, as it has a rugged beauty. It grows to about 15 feet and has a gnarled blackish trunk and branches. The stiff greyish-green leaves are spiralled around the branches and have a tufted appearance. The young growth is bright scarlet and very showy in spring. A striking parasite of the mistletoe family is common on *P. caffra*, but does it no harm and its red flowers often enhance the appearance of the tree.

The flower-heads are at their best around Christmas-time, blooming for a long period in midsummer. Each shallow, wide open flower-head measures about 3 inches across. The bracts are deep pink and the centre is filled with a mass of spiky white flowers. The flowers have a pungent, fruity odour that can be overpowering in a room. They were used in the design on our smaller silver coins.

P. caffra is available in the trade and is useful in gardens in the summer-rainfall area, where it is extremely drought-resistant, needing no attention once it is established. It will endure poor soil, but needs a warm, sheltered position, for severe cold will kill it in the young stages. It normally grows in acid soil, but sometimes grows in alkaline soil in nature. *P. caffra* grows wild on the hills around Johannesburg, from Bryanston northwards to Pretoria and as far west as Krugersdorp. It sometimes grows side by side with *P. roupelliae* and is also found on the lower slopes of the Drakensberg mountains in Natal.

All the Transvaal forms, formerly known as *P. multibracteata* and *P. rhodantha* and its varieties, are now regarded as forms of *P. caffra*, which is a huge, variable complex. Others, with underground stems and short trailing branches, also occur in the Transvaal and Natal, but have little horticultural merit.

P. gaguedi (formerly *P. abyssinica*) occurs from the eastern Transvaal escarpments to Abyssinia. It has silvery bracts surrounding the white flowers.

P. welwitschii (*P. hirta*) from 6 to 10 feet in height, has rather untidy cream flower-heads.

P. simplex is a dwarf 2-foot species with an underground rootstock which has a flower almost identical to that of *P. caffra*, but is smaller in proportion. It grows wild in the same area.

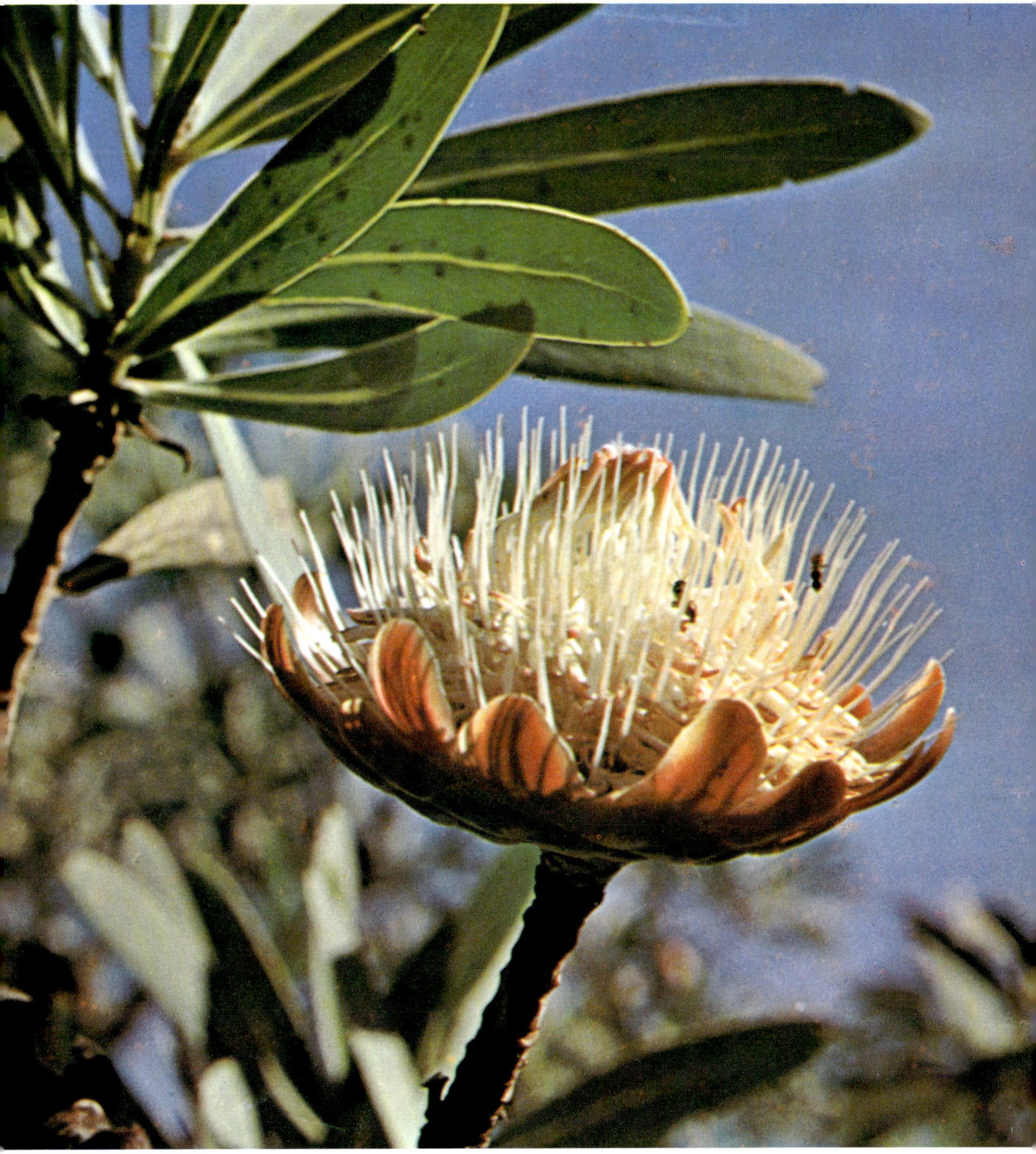

DRAKENSBERG PROTEA, ROUPELL'S PROTEA

Protea roupelliae

One of the showiest of the summer-rainfall Proteas, this is a large tree that grows to a height of 10 or 15 feet. It has a distinct trunk and the branches radiate out and up to form a rounded shape. The leaves are a silvery, bluish-green colour and cluster most luxuriantly near the ends of the thick branches. Each oblong leaf is about 4 to 6 inches long, pointed and covered with light hairs.

The flower-heads appear during midsummer at the tip of each branch, densely surrounded by leaves. At first glance, the flowers resemble *P. eximia* (formerly *P. latifolia*), for the bracts are widely spaced and spoon-shaped, narrowing near the base. But the cup is not so deep, nor the flower so large. The head measures about 4 inches across and is about 6 inches long. The central mass of hairs stands up in a blunt cone, coloured pink at the top and shading to yellow at the base. The pointed bracts are deep pink, almost red, shading to creamy yellow at the base.

While this is not one of the recommended Proteas for the average garden, it is excellent in a large garden in the summer-rainfall area, where it would be drought-resistant and require no watering for the six months of the dry season. It enjoys high rainfall during summer. It is fairly cold-resistant, yet sensitive to severe dry frost, in spite of the fact that it grows wild at high altitudes in the Drakensberg Mountains. It extends into Natal, Swaziland, the northern Cape and eastern Transvaal. It also occurs on the Magaliesberg Mountains and delightful groves may be seen on the open rocky hills near Krugersdorp and Magaliesburg in the Transvaal. *P. roupelliae* is available in the trade and grows slowly at first.

It grows in acid soil in nature, but has also been seen growing on dolomite, which is alkaline, so that it should be tolerant of a wide variety of garden soils.

The name *roupelliae* is now spelt in accordance with that of Mrs. Roupell, after whom it was named, although it was mistakenly published with a double 'p' when first named.

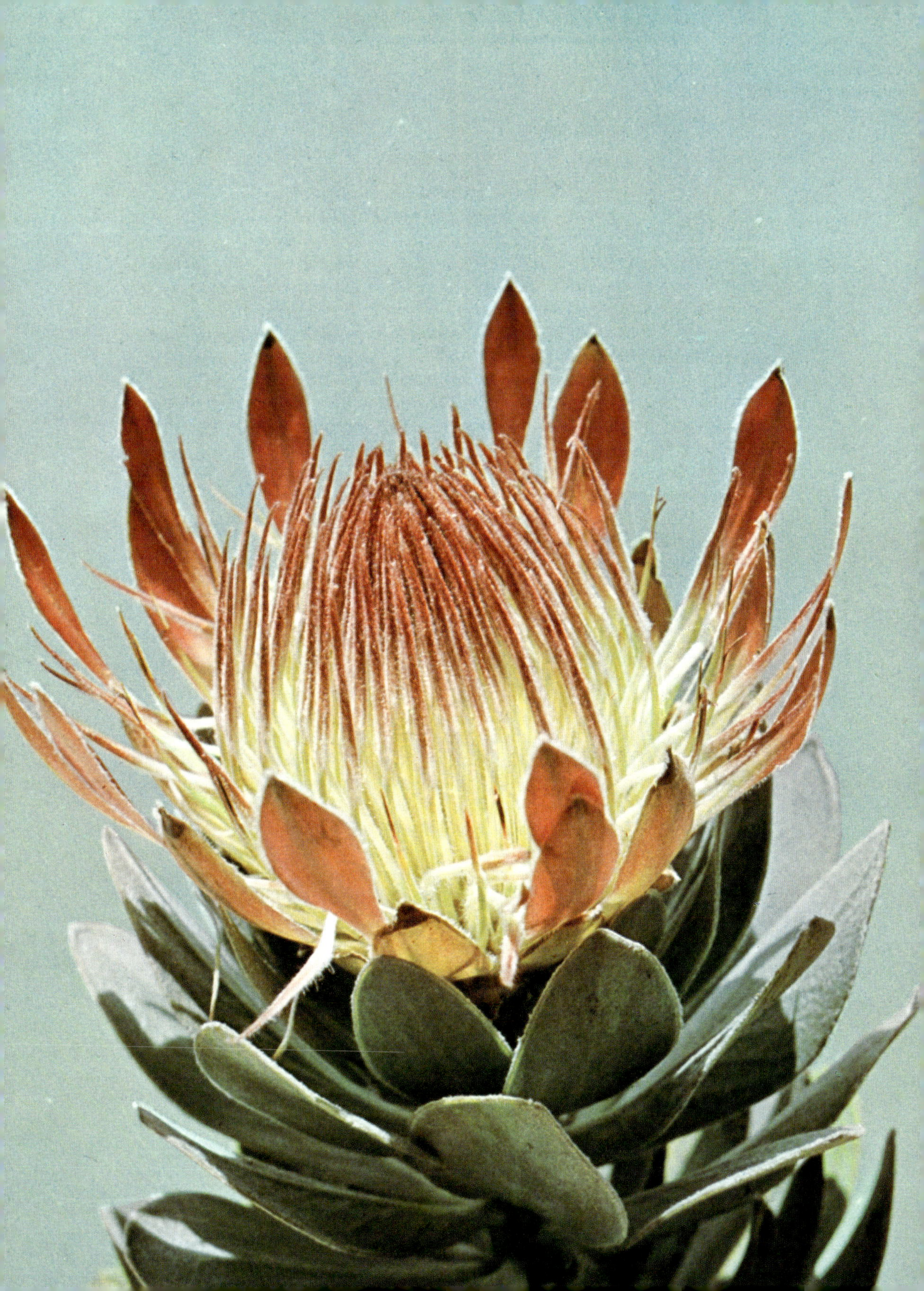

VELVET PROTEA
Protea rubropilosa

Rich golden-brown velvet buds, about three to four inches long and two inches thick, are the most attractive feature of this tree-like Protea. They retain their velvety texture for months, even when the flowers have been dried. If the buds are picked when they are tightly closed before drying, they will remain perfect for half a year or more, making excellent dried material for vases. Mature buds will open in water and will not dry out quite so neatly.

The bracts of this flower-head have a reddish-brown, velvety outer surface that make the buds so delightfully unusual. The inner colouring is a rich rose-pink. The flower-head opens fairly wide, measuring about three or four inches across, to reveal a wide high peak of bright pink flowers, tipped with brown, that remain fresh and colourful for about a fortnight. These flowers travel well out of water and bloom from August to October. They have a pungent, fruity scent that may be too strong for some tastes. The deep green leaves are smooth and very large, about six inches long and two inches wide. The tree itself is very gnarled, with corky bark, and grows to about 15 feet, although specimens up to 25 feet have been seen. (See opposite, above).

P. rubropilosa grows on the eastern Transvaal escarpment from Blyde River Canyon to the Wolkberg, at altitudes up to 6,500 feet. While it grows in stony grassland on the mountain tops and in well-drained, dry soil, it is definitely in the mist-belt and needs moisture. It grows in very acid soil and does not appear in the nearby dolomite (alkaline limestone) where *P. roupelliae* will grow. Seed germinates easily, but it is unlikely that this species will ever become popular in gardens.

P. comptonii, name after Professor R. H. Compton, who was the first to collect it, is a similar gnarled tree, but it occurs only on the rocky slopes of the mountains between Havelock Mine and Barberton. It has very long leaves, from five to ten inches in length, and one to two inches wide.

P. laetans. BLYDE PROTEA. The scientific name of this Protea means "happiness", as does the name Blyde, for it occurs only on a hill overlooking the resort at the Blyde River Canyon, eastern Transvaal. It was thus named by the botanist, Mrs. L. Davidson, in 1975, when she discovered that it differed from the other species in the area.

It differs from *P. caffra* in its flowering time, for it blooms from early autumn to mid-winter. The small tree is about the same size. It may be distinguished readily by its long, thick flower-stalk (stipe) which is covered with tiny silvery bracts. The deep rose-red flower-bracts are overlaid with silver hairs, as distinct from those of *P. rubropilosa*, which are covered with brown hairs. (See opposite, below).

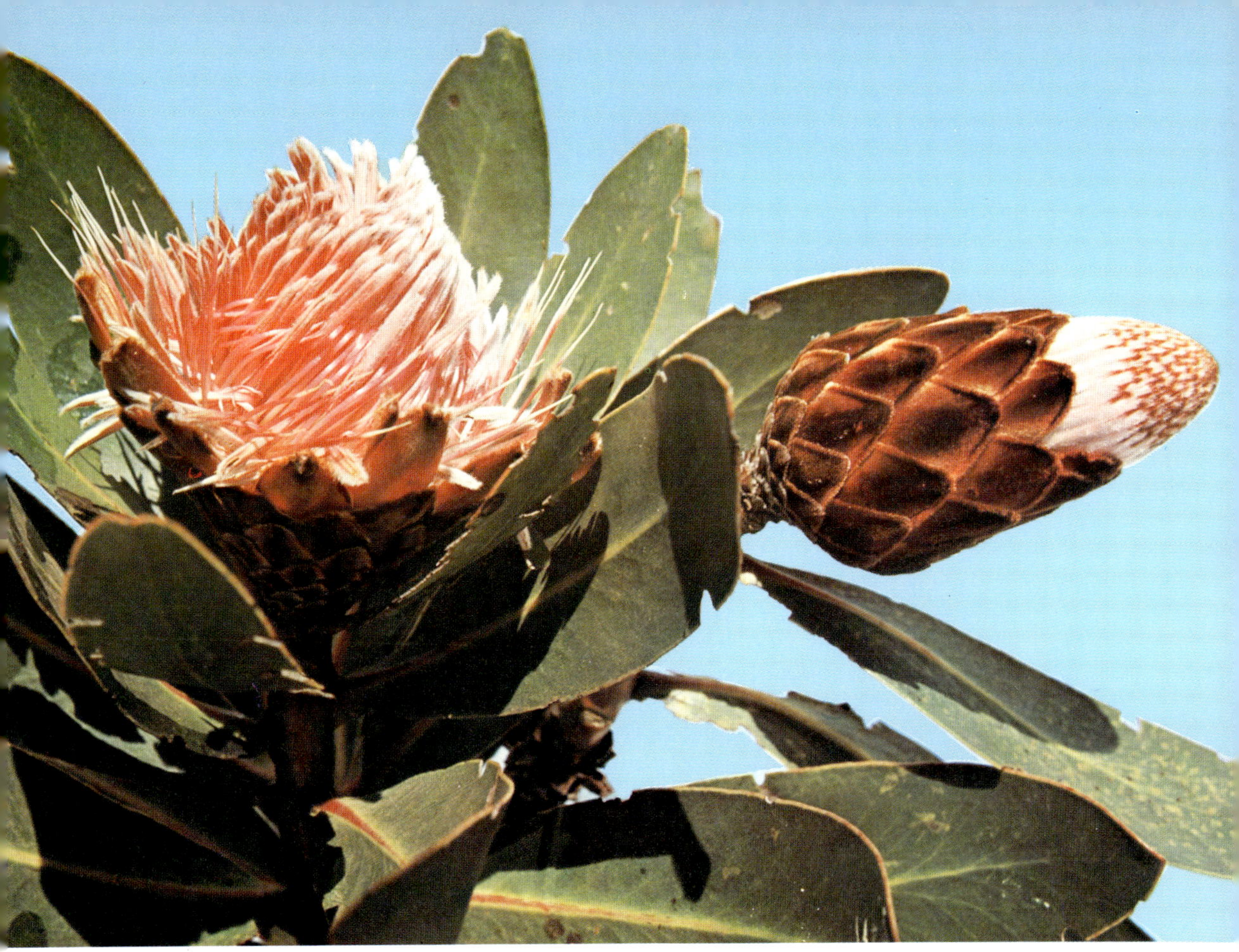

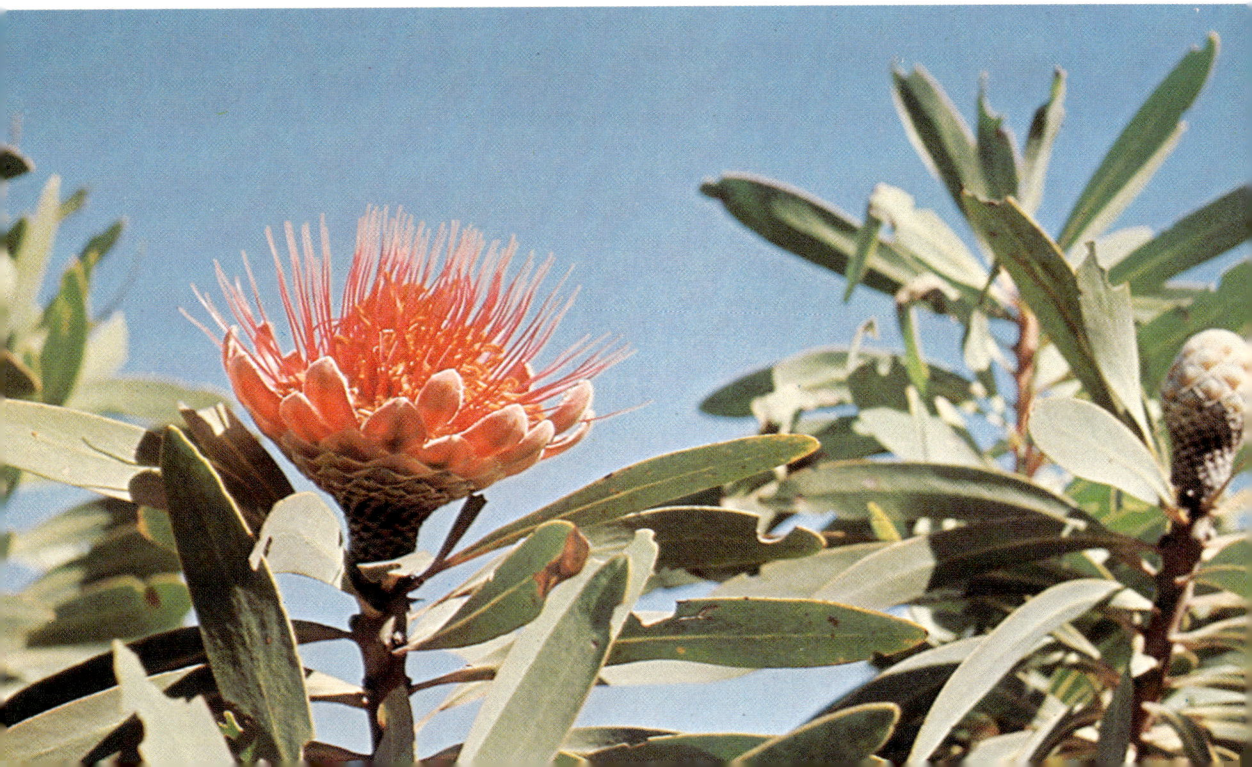

ON PROTEA HYBRIDS

While preferring the quality of species to hybrids, one must recognise the fact that hybrids appear naturally in cultivation as well as in nature. Some of them have an extraordinary beauty, combining the attractive features of both parents, so that one is tempted to reproduce them for horticultural purposes.

The flower pictured opposite is a natural hybrid with *Protea magnifica* and *P. obtusifolia* as its undoubted parents. The seed was originally obtained from Kirstenbosch by the author as *Protea magnifica*. When the bush grew rapidly to a height of 10 feet, it was obvious that this was not true to type, although the foliage resembled that of *P. magnifica*. When the gorgeous flowers appeared in the third season, it was a puzzle to decide what had happened, for they were not true Woolly-beards, even though the centre was filled with the customary mass of soft white woolly hairs. The central peak was red instead of black and the surrounding bracts were not as hairy as those of *P. magnifica*. The whole flower-head too, was smaller, measuring about 5 inches across. A study of the outside of the flower-head revealed *P. obtusifolia* as the other parent, for both its greenish colouring, tipped with red, and the formation of the scales was very similar to that species. The flowers appear from autumn to spring and it is one of the loveliest Proteas ever to be seen.

The only way to reproduce this bush would be by vegetative means, but several attempts have been made to grow it without success. Seed has so far been sterile. One is anxious to find a way to reproduce it for posterity before the plant dies. It is obvious, therefore, that in the field of hybrids, it would be a tremendous advantage to master the art of reproducing Proteas from cuttings. There is obviously a vast field to be explored in hybridising Proteas, but unless one conquers the vegetative method of reproduction, this will be a limited one. Coupled with the fact that there are so many natural variations in the wild, it seems that it will add further to the confusion of identifying the different species to breed hybrids indiscriminately.

VEGETATIVE REPRODUCTION

Since the first publication of this book, several nurserymen have propagated Proteas successfully by means of cuttings, with the help of mist-spray apparatus and heating of the soil by underground cable (to 24° C), as well as by means of air-layering, as illustrated by the photographs opposite.

The cuttings growing in the propagating bed under mist-spray are of *Protea magnifica, P. eximia, Leucadendron discolor, Leucospermum reflexum, Protea repens, P. cynaroides, Mimetes cucullatus, Protea neriifolia, P. punctata, Protea grandiceps, Leucospermum praecox* and *L. tottum*. The air-layering is on *Leucospermum reflexum*.

Several named cultivars have been produced by growers since 1975 and have proved popular cut flowers for export (see page 222).

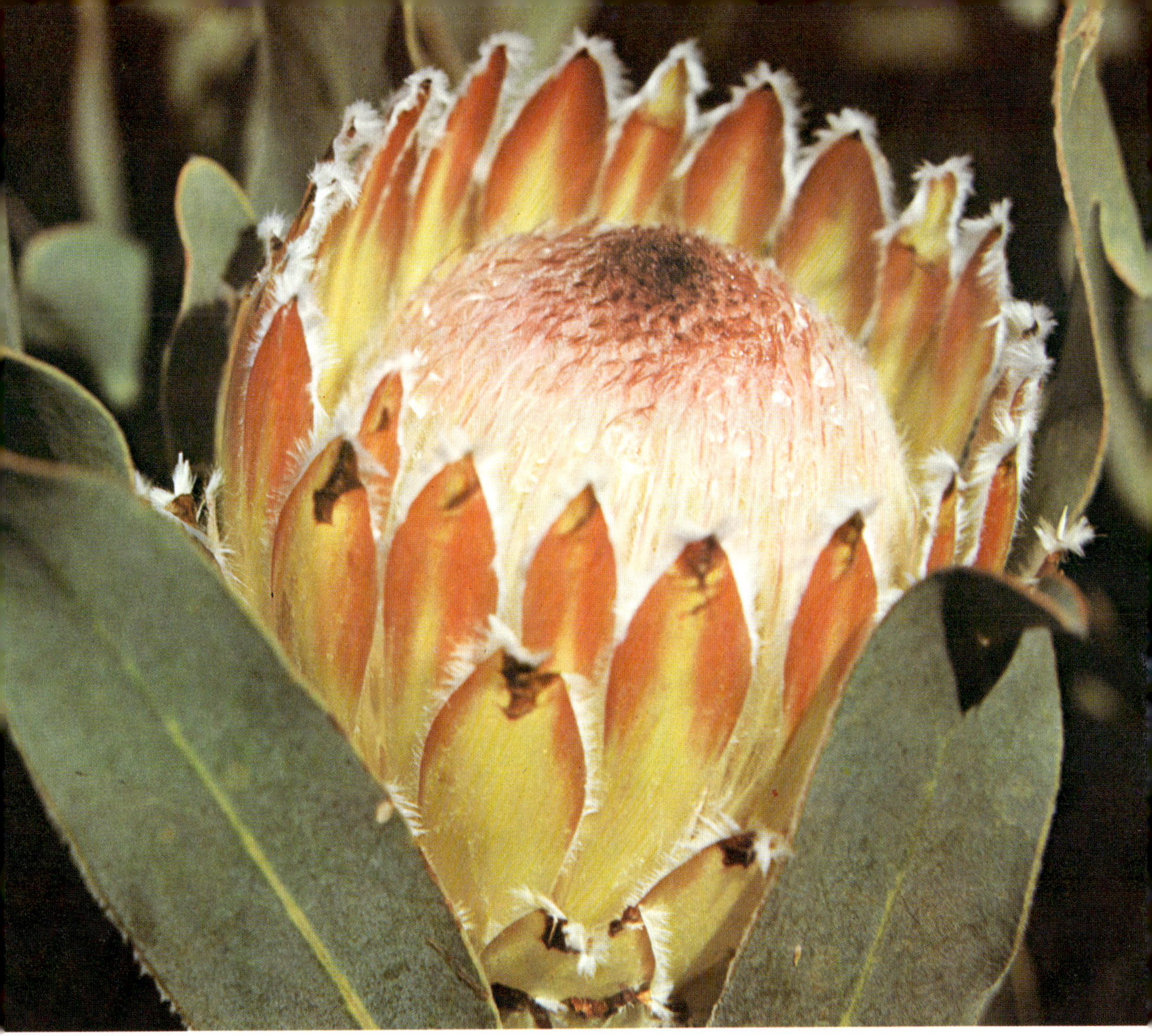

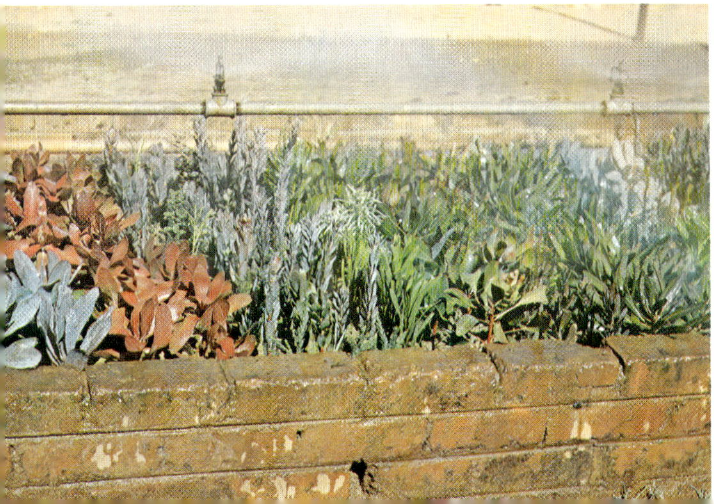

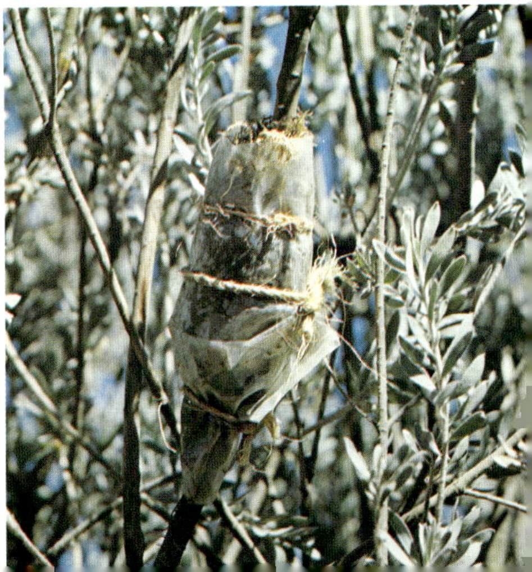

LADISMITH PROTEA
Protea aristata

This is one of the newest Proteas to excite the attention of horticulturists. It is remarkable for its fine pine-needle-like leaves that grow up to 3 inches long and clothe the branches with a fringe of soft fresh green. The flower itself is lovely and an exquisite glowing deep pink, almost red, when fresh.

The flower-head is about 4 or 5 inches across and the pointed, brown-tinged bud opens into a cup shape. The centre is filled with a cone of deep pink, hairy flowers. The flower-heads are borne at the tips of the graceful branches and bloom late in the year, in November, December and January. The bush grows to a height of about 5 or 6 feet, with an open branching habit and numerous short branches growing upright.

This is a slow-growing plant that is hardy to cold weather. It grows in rocky places in high mountains near the snow-line and is rare in nature. It was first discovered by T. P. Stokoe in December 1937 in the Kleinswartberg range of mountains in the south-western Cape, where there were only a few plants. It was grown from seed for the first time by Kirstenbosch in 1960 and the plants commenced flowering in November, 1964. Despite the fact that a handful of plants exist in nature, seed gathered by a nurseryman near Ladismith, Cape, have germinated in large numbers, so that plants are now not only obtainable in the trade, but it may become possible to reintroduce more plants into their natural environment.

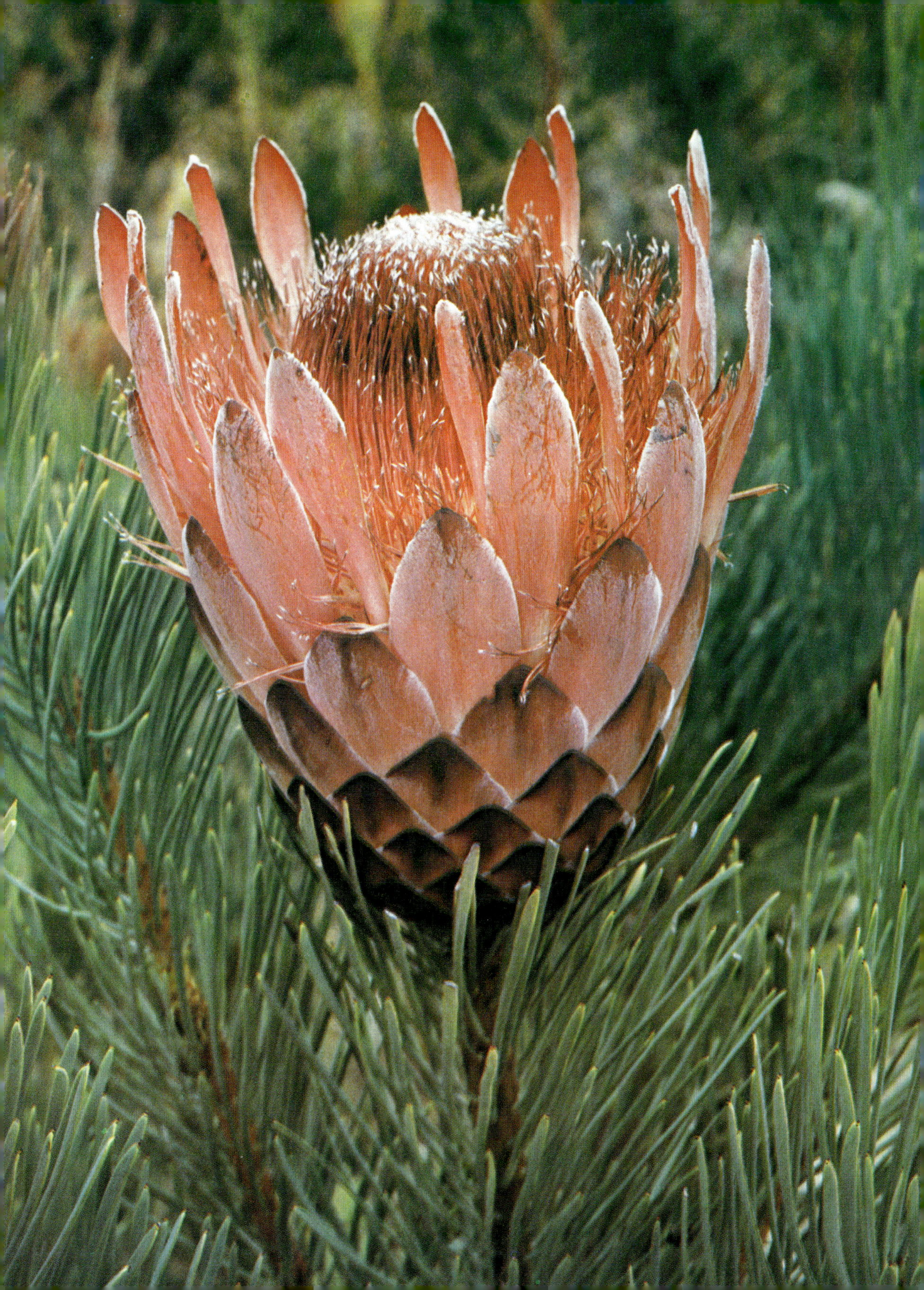

LONG-LEAVED MOUNTAIN ROSE
Protea pityphylla

A rare Protea in nature and in cultivation, the long-leaved Mountain Rose has a distinctive charm that makes it a desirable acquisition. The ruby-crimson flowers droop gracefully from the tips of the branches like those of *P. nana*, but they are larger —up to 2½ inches across, and open into a more shallow widespread bowl. The inner colouring is attractive, shading from red at the tips to cream at the centre, while the boss of hairy flowers at the centre has a golden-tan sheen. There is also a spreading form with greenish flowers that lies almost flat on the ground. The flowers appear in August and September.

Whereas the needle-shaped leaves of *P. nana* are short and cluster closely up the stems, those of *P. pityphylla* are long, measuring up to 3 or 3½ inches in length. The flower seems to spring away from a tuft of leaves, which stream backwards like the crest of a bird. The leaves are sparser than those of *P. nana* but grow along the length of the stems in the same way.

The whole bush is fairly open in habit, unlike the dense growth of *P. nana*, and makes a dainty shrub for a rockery. The long-leaved Mountain Rose has only a few long branches, about 6 feet in length, and very few side shoots. They curve outwards gracefully, borne down by the weight of the flowers which form on the tips, sometimes even trailing on the ground.

P. pityphylla grows wild in a restricted area of the Ceres mountains of the south-western Cape. It is fairly cold-resistant.

Protea effusa (P. marlothii) is a similar rare Protea, which comes from the Ceres district. The flower-heads also nod like those of the Mountain Roses, but they are larger than either, being nearly 3 inches across. The shallow bowl-shaped heads are a satiny crimson, shading to greenish-cream at the base of the bracts. The leaves make this species easily distinguishable from *P. nana* and *P. pityphylla* as they are not needle-shaped, but almost half an inch wide at the broadest part and 2 to 3 inches long, narrowing at the stem end. The bush grows to a height of 3 to 4 feet and flowers in spring.

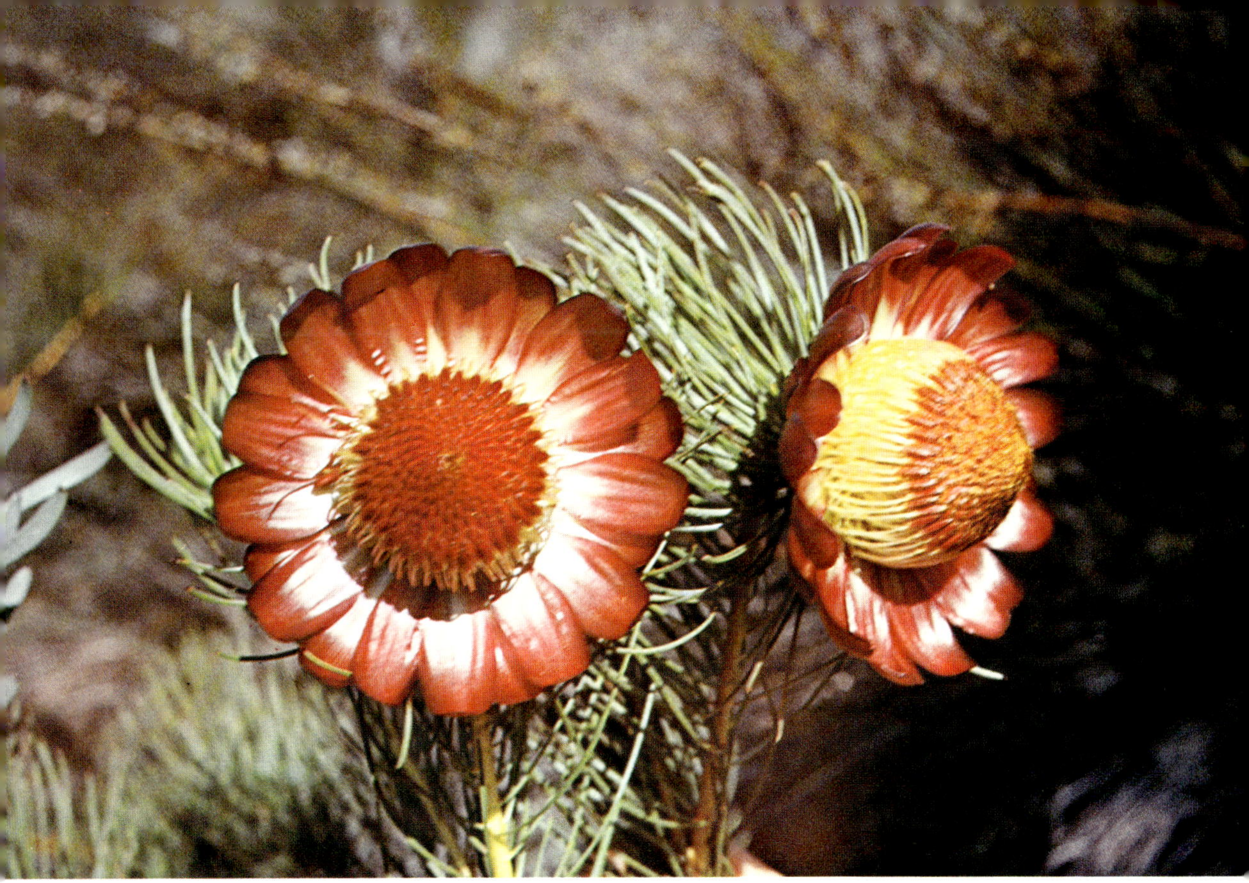

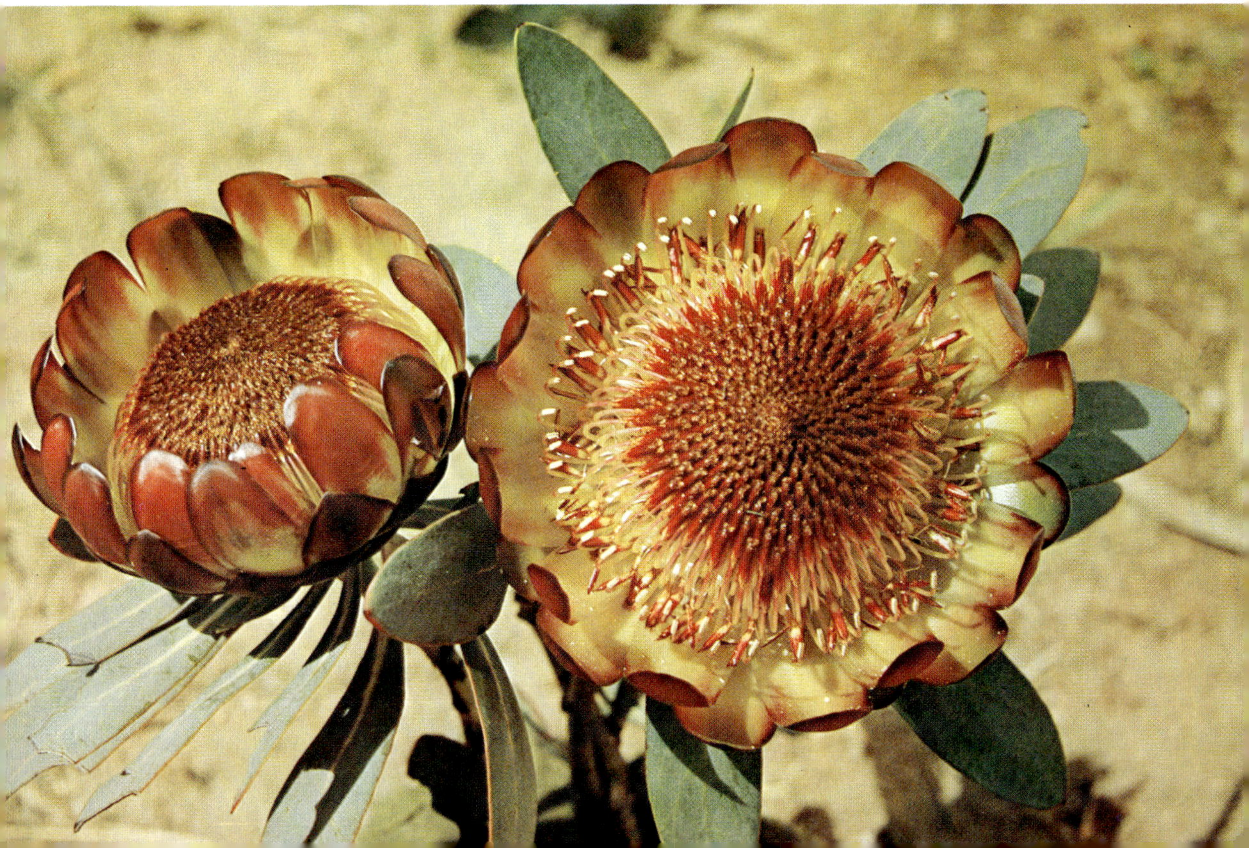

CEDARBERG PROTEA

Protea acuminata (P. cedromontana)

A charming small Protea, this is worth cultivation, for the burgundy-coloured flowers are unusual and dainty. This is a small erect shrub, growing to about three or four feet, so that it could be included in a small garden. It might make an interesting spring-flowering evergreen for a tub on a sunny patio. The bush is not very dense and has an open habit.

Essentially a neat flower, the heads are borne upright and measure about $1\frac{1}{2}$ to 2 inches across, while the glossy red bracts curve inwards gracefully. The central mass of real flowers is burgundy and the yellow tips stand out well against the dark background. The dainty bluish-leaves are narrow, growing up to about $3\frac{1}{2}$ inches in length.

Interesting because it grows wild in the Cedarberg mountains of the south-western Cape, right up to the snowline, but at lower elevations than the Sneeublom (*P. cryophila*), this is cold-resistant. It also grows near Villiersdorp. This pretty little Protea is obtainable in the trade and is a plant for the collector.

Protea canaliculata (P. harmeri) is very similar to *P. acuminata*. The flower appears to be identical in size and colour. It forms a stocky hard bush with short stems, very compact and rounded, which grows to four feet. The leaves are a slightly silvery green and about half the length of those of *P. acuminata*. *P. canaliculata* grows fairly profusely in the Swartberg Range of south-western Cape mountains, near the top of the pass.

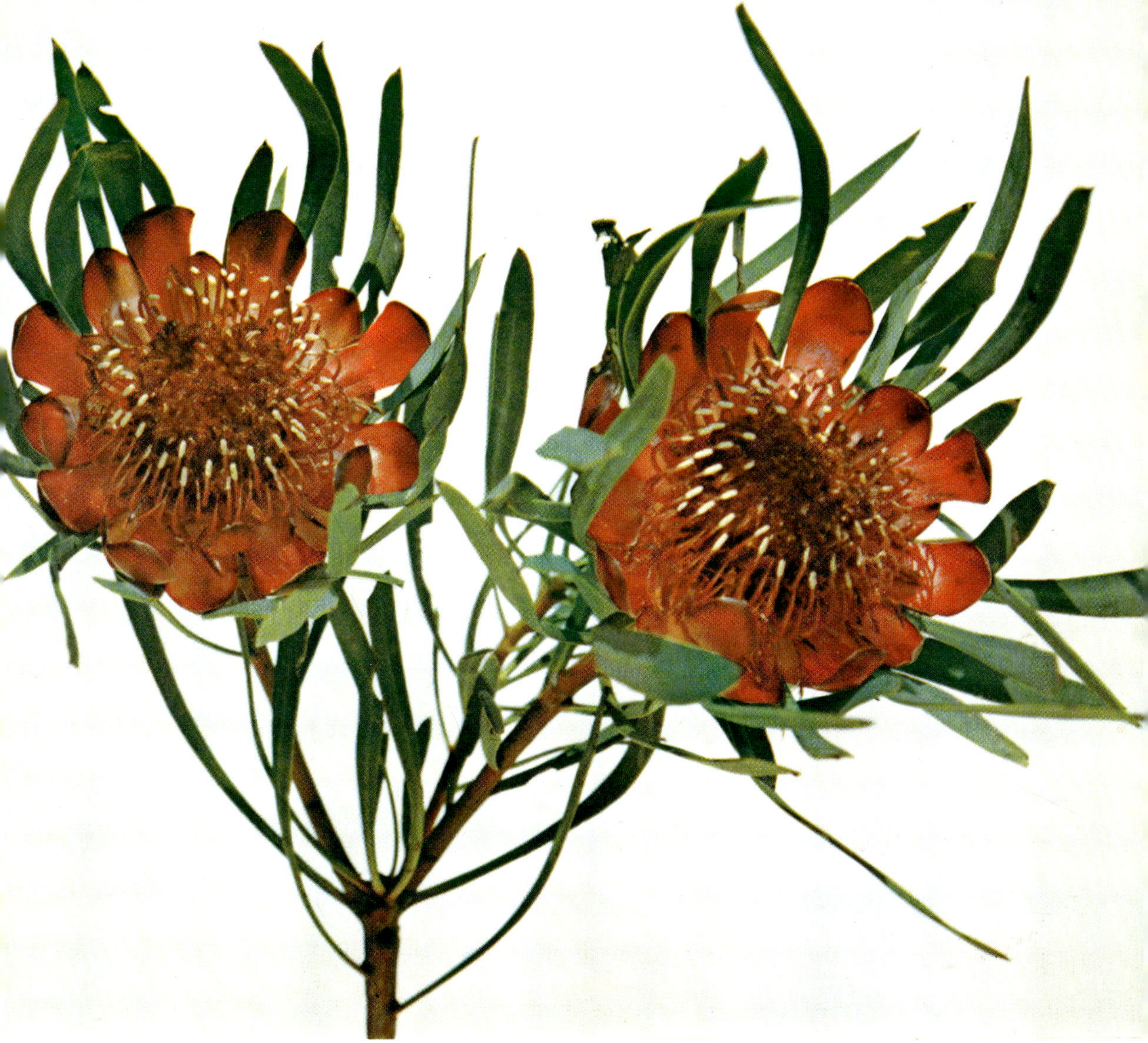

NODDING PROTEA
Protea pendula

This lovely little Protea was in danger of extinction, but seed was collected and it is now obtainable in the trade. It has small bowl-shaped flower-heads, measuring up to about three inches across, with a distinct silky covering of silvery hairs on the outside of the bracts, giving it a pleasant sheen. The bracts are a coppery red, sometimes tinged green, and the hairy flowers at the centre are tipped with golden brown. The heads hang down gracefully from the stiff branches in the spring.

The bush itself is erect, growing to a height of four feet and the stems are covered with oblong, greyish leaves. These twist and grow upwards, standing away from the heads in the same way as those of *P. pityphylla*. They are about two to three inches long and $\frac{1}{4}$ inch wide.

P. pendula grows wild in a restricted area in the mountains of Tulbagh and near Ceres in the Cedarberg, where snow falls in winter. It should be cultivated in the same way as most Proteas from the south-western Cape and should make an attractive evergreen shrub for a large tub. It should be compared to *P. witzenbergiana*, which differs from it in having thin, needle-like leaves and is more of a low, spreading bush.

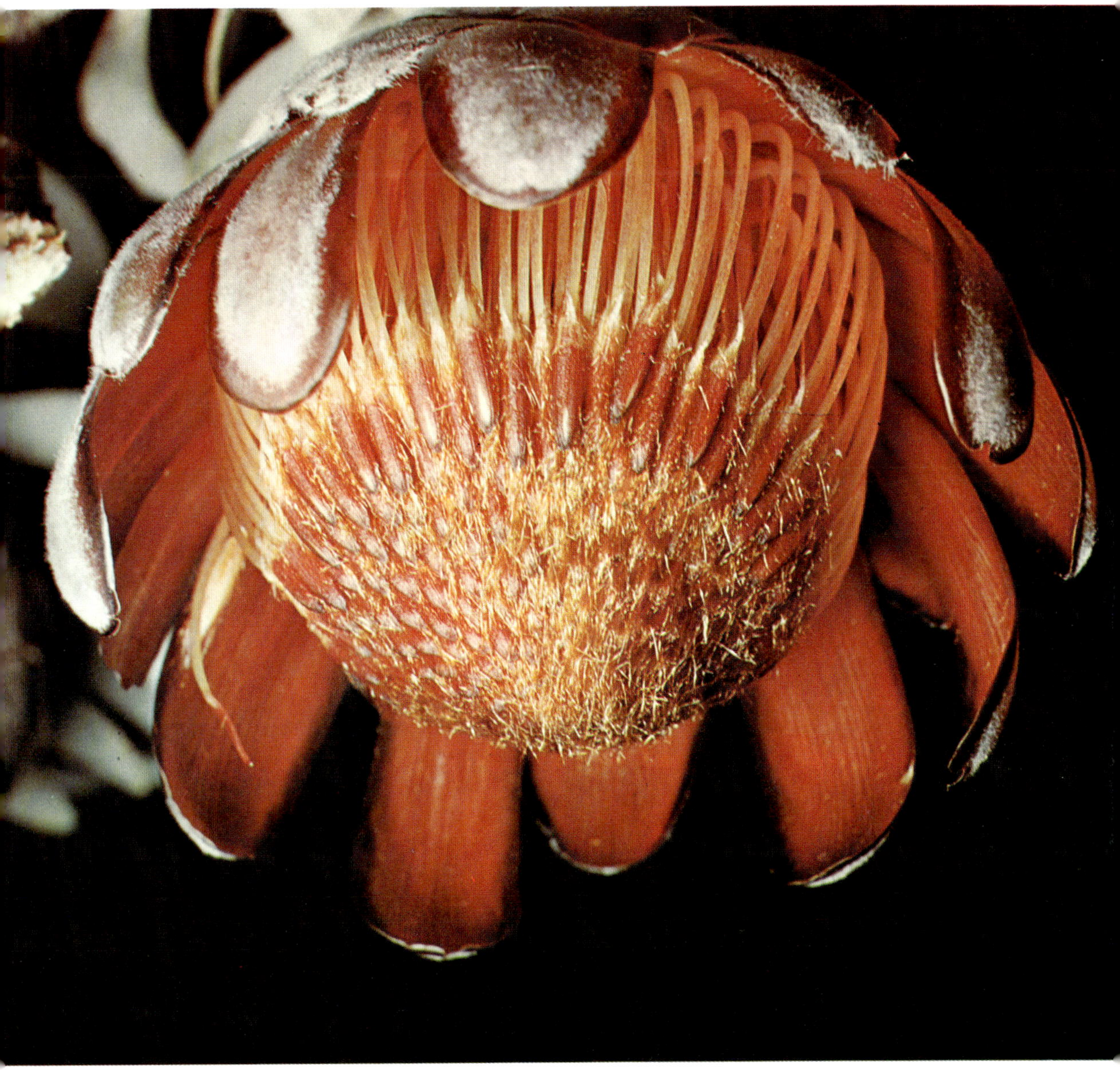

SULPHUR-COLOURED PROTEA
Protea sulphurea

This rare Protea has exquisite colouring and a fresh appeal. The blooms travel well out of water and this would be worth cultivating if seed were available. The flower-heads measure about 3 to $3\frac{1}{2}$ inches across and open wide to reveal the yellow bracts, shading into rose, and the perfectly shaped beehive of hairy flowers at the centre, which are sulphur-yellow, tipped with golden-tan. The reverse side of the flower-head is attractive, each yellow overlapping bract being neatly edged with rose. This may be seen in the frontispiece illustration where a half-open flower of *P. sulphurea* lies on the table. The flowers have a strong smell of honey and the heavy pollen comes off easily.

This is a large spreading shrub which grows to a height of about two feet, but spreads over an area of about ten feet. The flowers hang downwards with their faces to the ground so that they are not very showy in the garden. The small, tough, bluish-green leaves are broader near the tips.

It grows wild in the high peaked mountains in the drier districts of the south-western Cape, in the vicinity of Robertson, Villiersdorp, Montague, Touw's River, Laingsburg and Ladismith, as well as in Namaqualand on the Kamiesberg. It grows near *P. acuminata* and *P. grandiceps*. Although it does not receive much rainfall in nature, cloud mists provide moisture and the plant would need regular watering in cultivation as well as well-drained, acid soil.

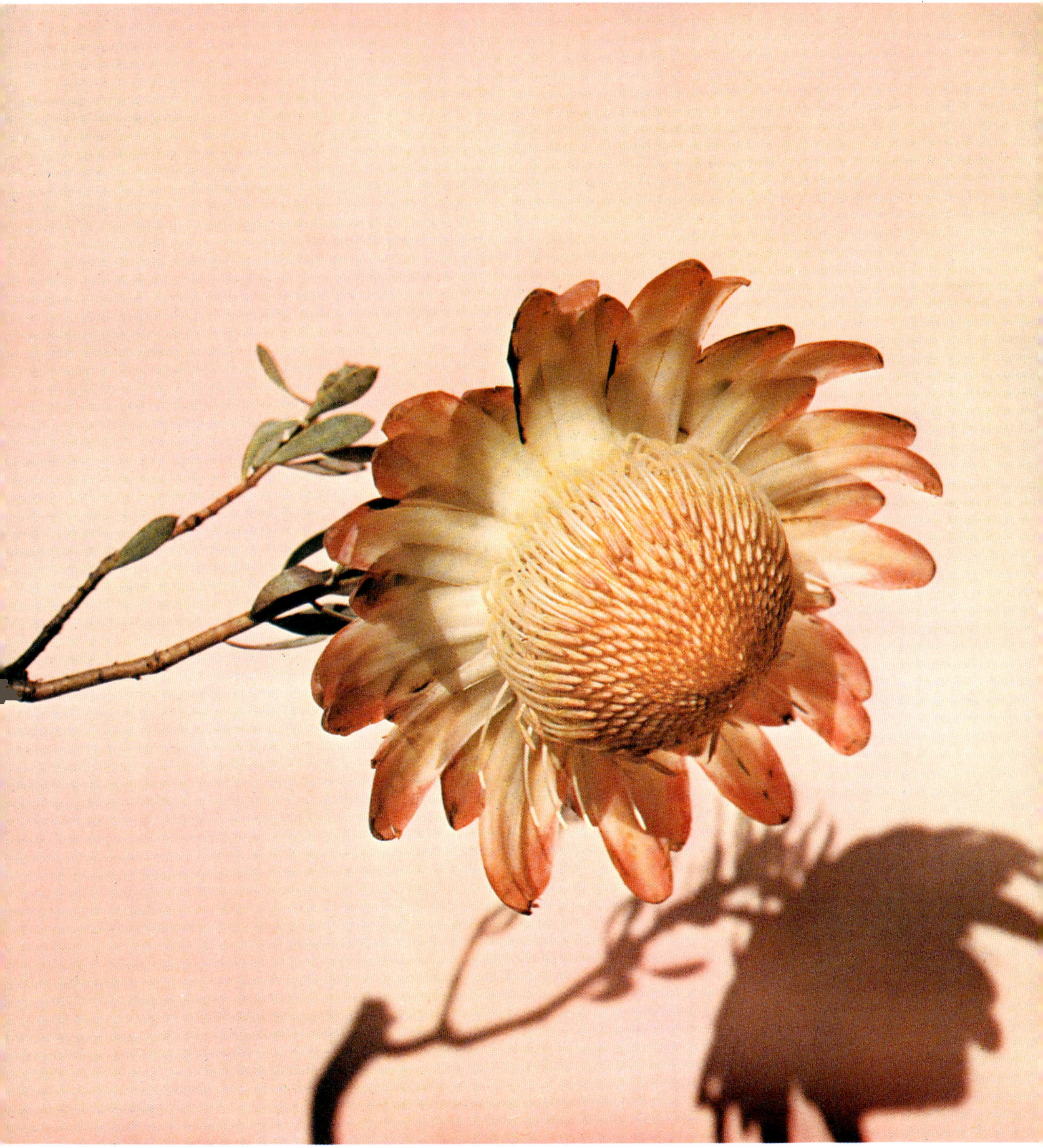

STEM-CLASPED PROTEA, IVY-LEAVED PROTEA

Protea amplexicaulis

It seems strange to appreciate a Protea more for its foliage than for its flowers, but the trailing branches of *P. amplexicaulis*, covered with small, bluish-green triangular leaves that are arranged artistically around the stems, have an ivy-like charm and are much prized by people who love to arrange flowers. The leaves are edged with red and the young leaves are generally blood-red. They last well in the vase.

This is a sprawling plant that covers the ground with foliage, its graceful stems spreading out over an area of about 2 feet. It only grows up to one or two feet in height, forming a rounded plant, and makes a good subject for a rockery or sloping bank. It could make an interesting evergreen pot plant in a sunny situation.

The flowers develop underneath the stems, which must be lifted to examine them. These have a strange beauty, being a velvety purplish black, with dark centres sprinkled with yellow pollen, and look dramatic if they are picked and placed in a shallow white bowl. The flower-heads measure about $2\frac{1}{2}$ to 3 inches across and the bracts curve inwards in a bowl shape. The flowers first appear in early autumn and continue flowering throughout winter and spring.

This species is obtainable in the trade and grows easily. It must have regular watering, especially in winter, and will grow in sandy soil. It stands up well to cold weather. *P. amplexicaulis* grows wild on the sandstone hillsides of the south-western Cape in the Caledon, Tulbagh, Ceres, Worcester, Paarl and Bredasdorp areas. It grows at altitudes up to 4,000 feet and is frequently found growing near *P. magnifica*.

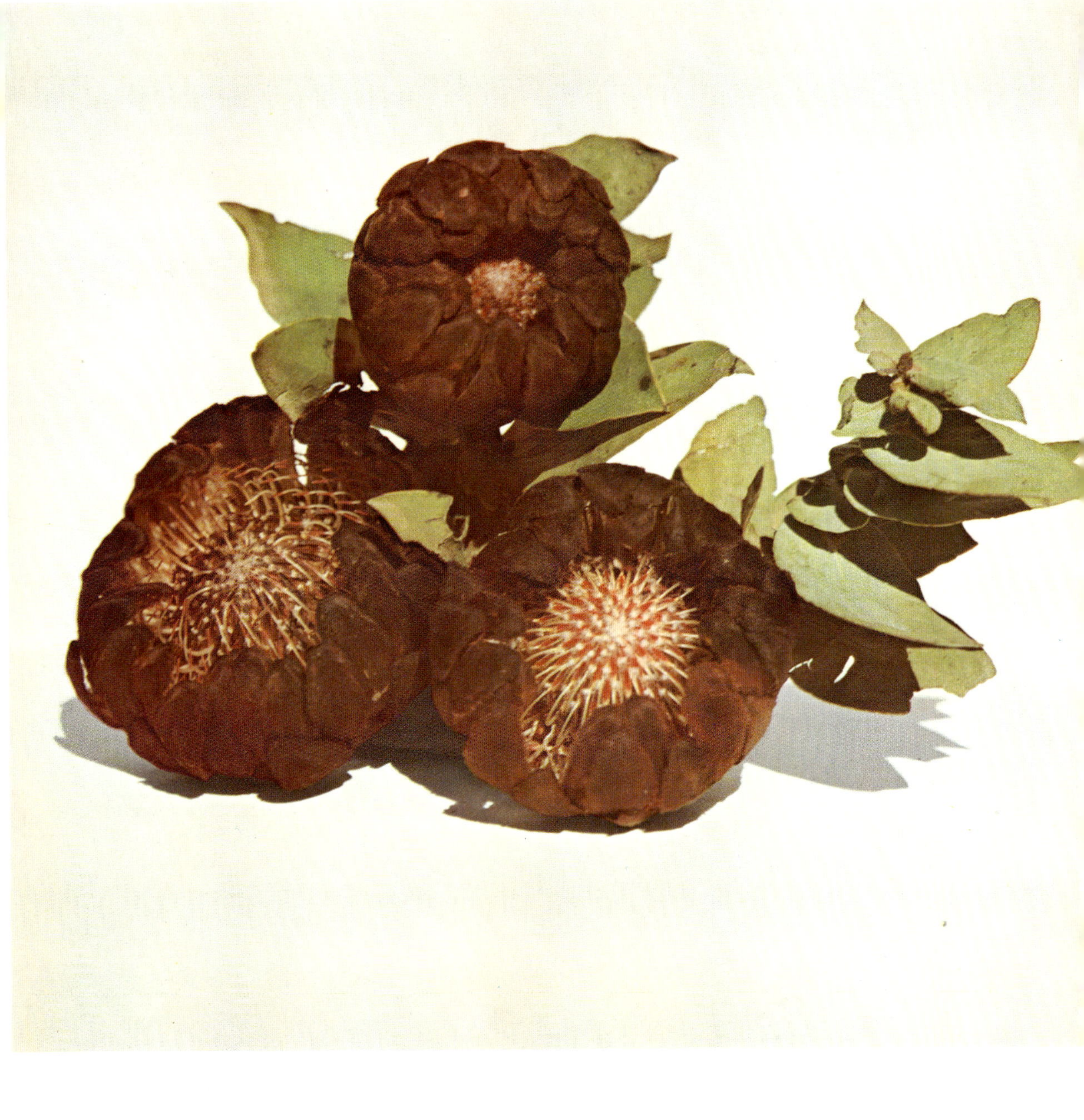

NEEDLE-LEAVED PROTEA

Protea subulifolia (P. acerosa)

This strange Protea has an underground root and main branches which lie half buried in the sand. The flowers lie just above the surface, being thickly clustered along the length of the thick old branches. Short side branches, twelve to eighteen inches long, also emerge erect from the main stem and are lightly covered with soft needle-like leaves. Other Proteas with "needle-like" leaves include *P. pityphylla, P. nana, P. witzenbergiana* and *P. aristata.*

The flower-heads seem to be grouped at the centre of the plant and are pretty when one picks and examines them closely, but have little horticultural value. Each flower-head measures about $1\frac{1}{2}$ to 2 inches across and has a rounded shape, with the "old rose" bracts curving over and almost enclosing the cluster of pollen-tipped flowers at the centre.

This dwarf plant grows at low elevations in acid, sandy soil near Bot River and Betty's Bay in the Cape. It is chiefly of botanical interest and belongs to a group of Proteas with underground main stems, which include *P. acaulos, P. amplexicaulis, P. aspera, P. cordata, P. cryophila, P. humiflora, P. lorea* and *P. scabra.*

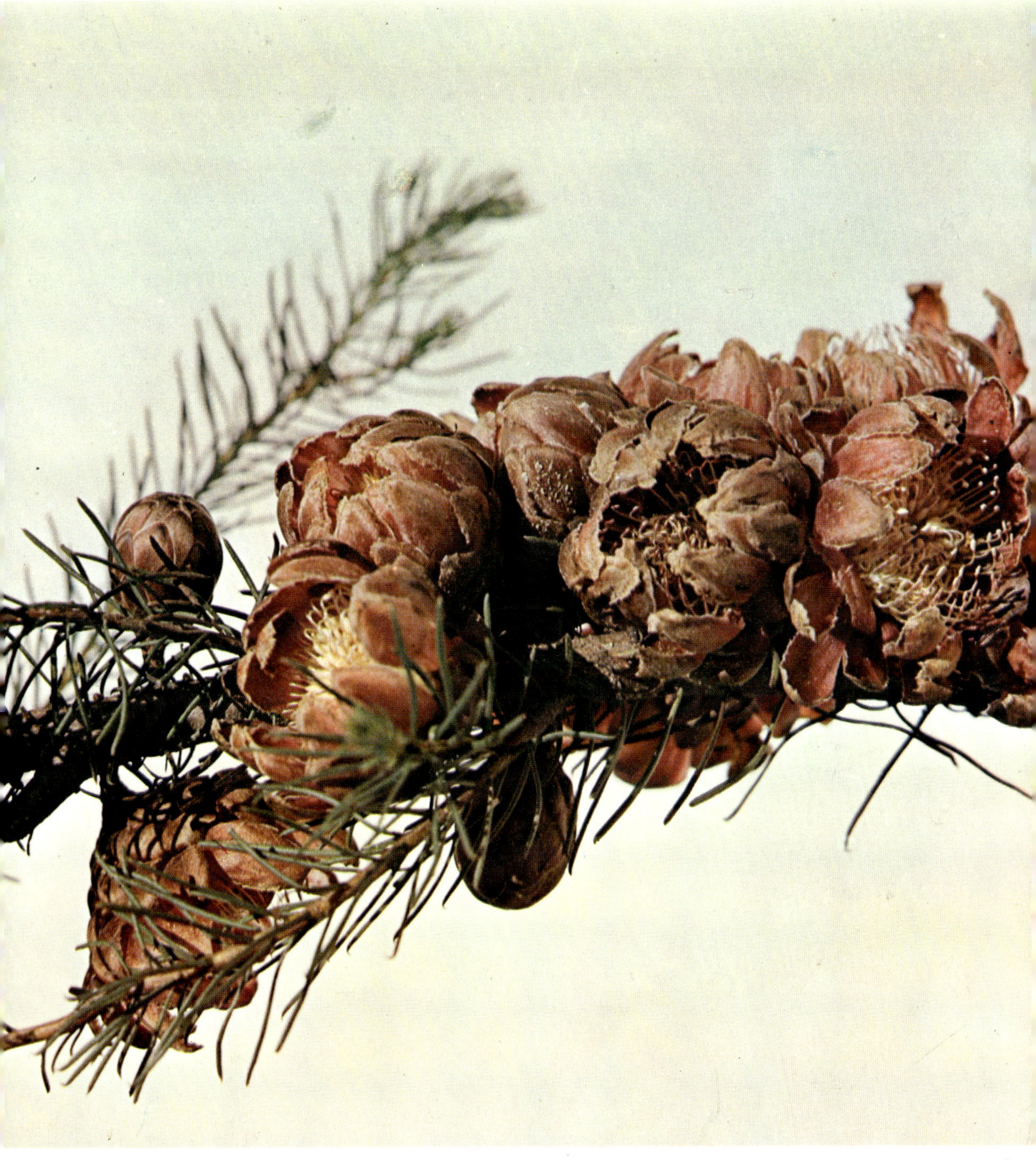

LONG-BUD PROTEA
Protea aurea (*P. longiflora*)

Although it is by no means one of the most desirable Proteas, *P. aurea* can be relied upon to flower profusely, especially around midsummer, when there are few other Proteas in bloom. Even if it were not grown for its flowers, it would make a useful evergreen background shrub or a screening plant, as it forms a dense bush, growing to a height of 10 feet and measuring at least 6 feet in diameter.

The flowers may be recognised most easily when they are in bud, for they are extremely slim, less than an inch across and about 4 to 6 inches long. The best colour form is a deep pink, almost red, but the white or cream form is seen most commonly. Both colour forms appear in the illustration. The hairy flowers that are usually massed in the centre of most Proteas are very sparse in *P. aurea*, so that the cup appears empty at first. The flowers do not last long and should be picked in bud. They will open in the vase and generally open wide quite suddenly. They flatten outwards fairly rapidly and the dried flowers stand up in a tuft in the centre. Dead flower-heads should be cut off as soon as they fade, otherwise the bush looks very untidy.

The deep green leaves of *P. aurea* are fairly short and broad, a little more than $\frac{1}{2}$ an inch in diameter and about 2 or 3 inches long. The numerous branching stems, covered with foliage, give it a dense appearance.

P. aurea grows easily in average well-drained garden soil, needing sunshine and regular water, but it is quite drought-resistant. It seems to withstand cold quite well and is freely obtainable in the trade. It grows wild in the south-western Cape mountains, near Worcester, Robertson, Swellendam and Mossel Bay.

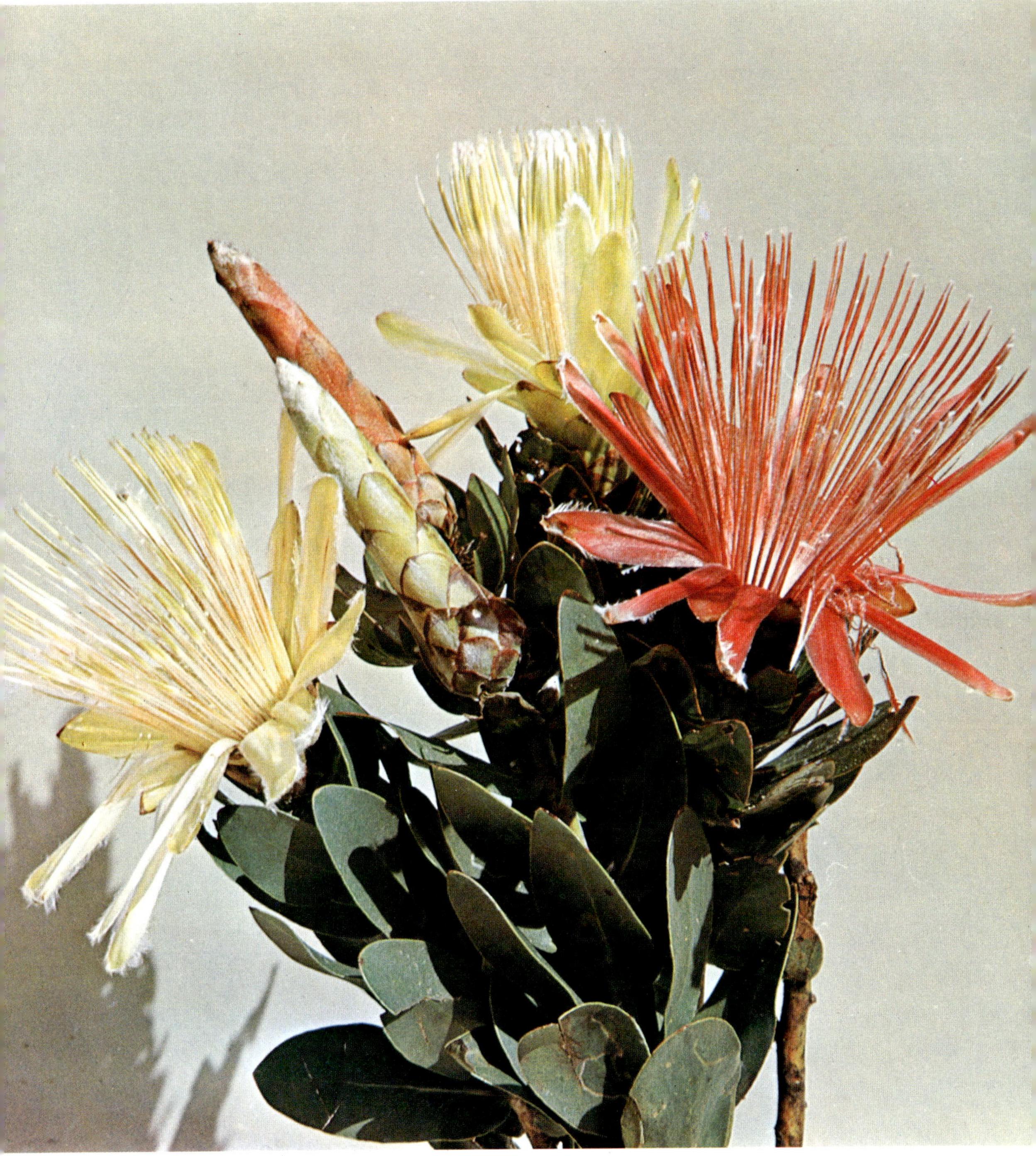

LONG-LEAVED PROTEA
Protea longifolia

Plants with white flowers are not generally popular, but the exquisitely formed flower-head of *P. longifolia* makes it a desirable acquisition for the vase. Its light colour is particularly good in a mixed bowl, where it offsets the deeper tones and has a brightening effect. It blooms for a long period, from autumn through winter until late spring.

The flower-head is about 5 or 6 inches long and 4 or 5 inches across at the widest part. The pointed bracts are generally cream with a touch of pale green or yellow and may sometimes be tinged with pink. They enclose a fluffy white mass of hairy flowers that culminate in an elegant peak tipped with black, a dramatic feature that is similar to that of *P. magnifica*, but the "spot" is more pointed.

P. longifolia is recognised easily even when it is not in flower by its very long narrow leaves that measure about 6 or 7 inches in length and less than $\frac{1}{2}$ an inch across. They point upwards along the stems, half enclosing the flower-heads at the tips of the branches.

This is a low spreading shrub that would fit into a small garden as it grows from 3 to 5 feet in height. It will clothe a sloping bank or drape gracefully over large rocks. It thrives easily in average well-drained soil, but grows in sandy soil in nature, on hillsides in the Caledon division of the south-western Cape.

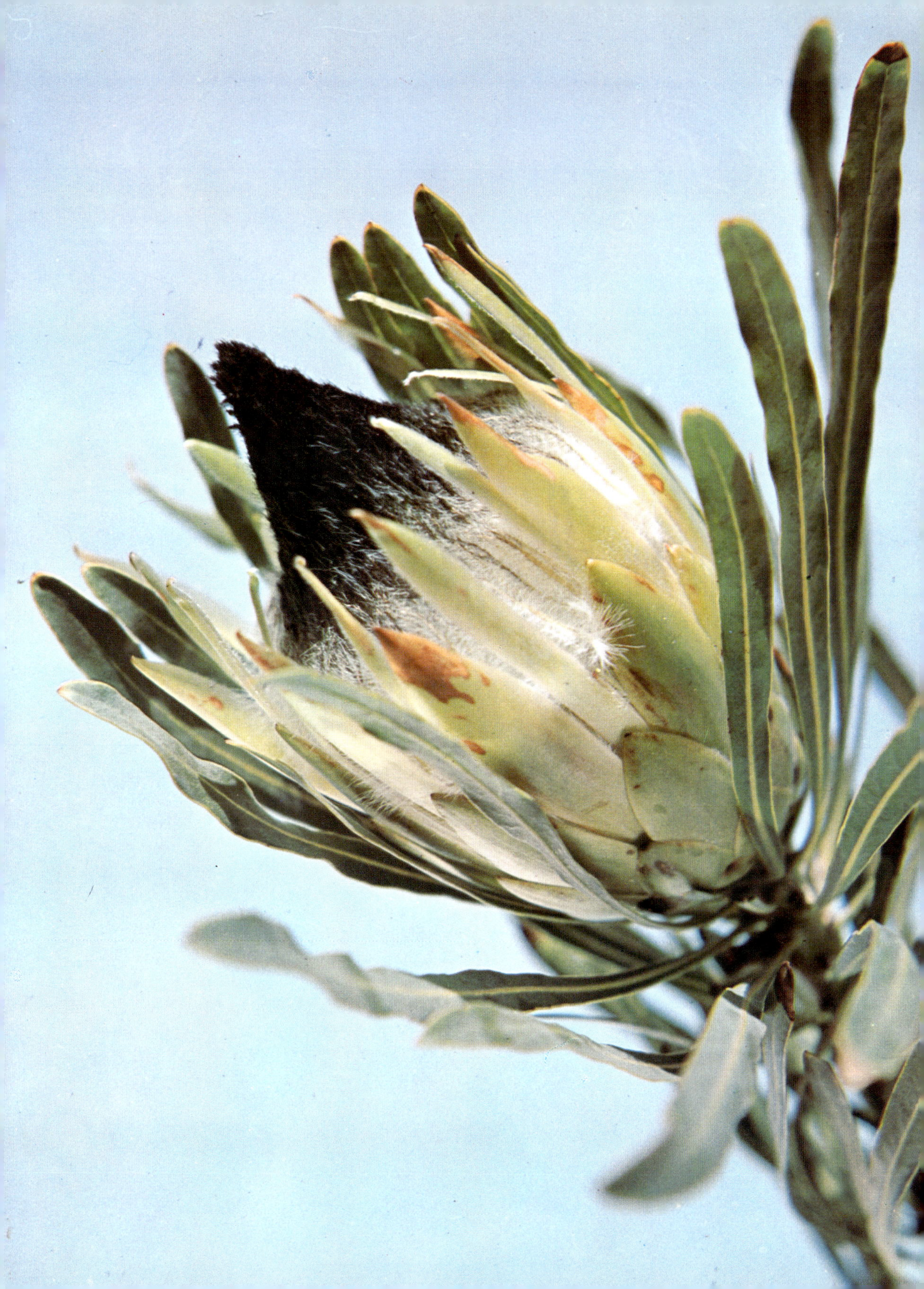

BOT RIVER PROTEA
Protea compacta

One of the earliest Proteas to herald the winter season, this familiar flower seems to blossom in all the florists and roadside stalls in May. It is a characteristic lovely clear pink that has been adopted as the fashion colour called "Protea Pink." The flowering season lasts throughout autumn, winter and spring and well into the summer. This prolific bloomer will provide fortnightly bunches of flowers all through winter once the bush is about five or six years old, but commences flowering in the third year from seed.

Each flower-head, which is borne at the top of the branch, measures about $4\frac{1}{2}$ inches in length and 4 inches wide. The broad pink velvety bracts curve inwards near the top half enclosing the central mass of soft pink flowers, and this is a distinguishing character. The oval, light green, stemless leaves clasp the branches and overlap upwards. Unfortunately, these tend to blacken early and do not last as long in the vase as do the flowers. They can, however, be stripped if they are needed for a long period.

The bush grows up to 10 feet in height, developing long slender branches that tend to make the bush fall sideways as it grows. One should stake the young plant and pinch it back or cut the flowers regularly with long stems in order to try and keep it from becoming top heavy. Too much moisture encourages tall growth and the plants tend to fall when the soil is soft and wet. Groups of plants growing in nature seem to support one another.

P. compacta grows easily in average well-drained soil and seems to resist cold fairly well. It needs regular watering during the winter months. It is common in the Caledon area of the south-western Cape, near Kleinmond and Bot River.

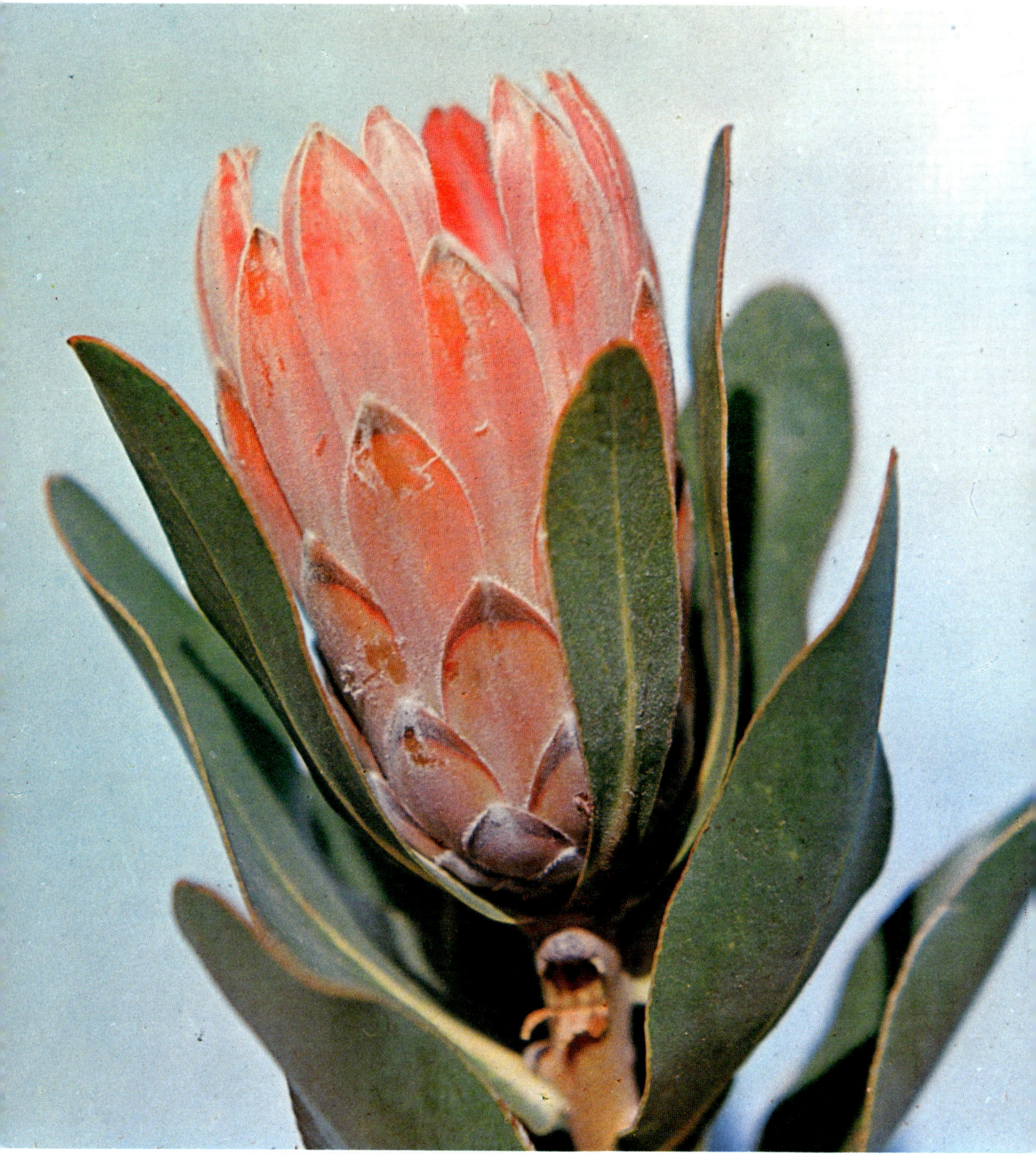

BLACK-BEARDED PROTEA
Protea lepidocarpodendron

The heavy black fur on the tips of the bracts of this Protea make it a dramatic flower as well as an easily recognisable species. The inner bracts, which form a striking contrast to the black, are cream or palest green, while the lowest, overlapping outer bracts are brown. The comparatively slim flower-head, for this does not open wide, grows up to about five inches in length. This species flowers over a very long period, starting in autumn and continuing throughout winter and spring until early summer.

The bush grows up to a height of about four to six feet and the branches are covered with deep green leaves which are about $4\frac{1}{2}$ inches long and $\frac{3}{4}$ of an inch wide.

P. lepidocarpodendron grows easily in gardens in a wide variety of soils and climates. It enjoys well-drained soil and seems to withstand drought quite easily, but will respond to regular watering throughout the year. This is a wind-resistant plant which is valuable for sea-coast gardens. It is found growing wild in a wide variety of conditions in the south-western Cape, extending from the windy dry areas near Port Elizabeth to the Clanwilliam Division. It is also common on the Cape Peninsula, where rain falls in winter and spring, but the sea and cloud mists provide summer moisture.

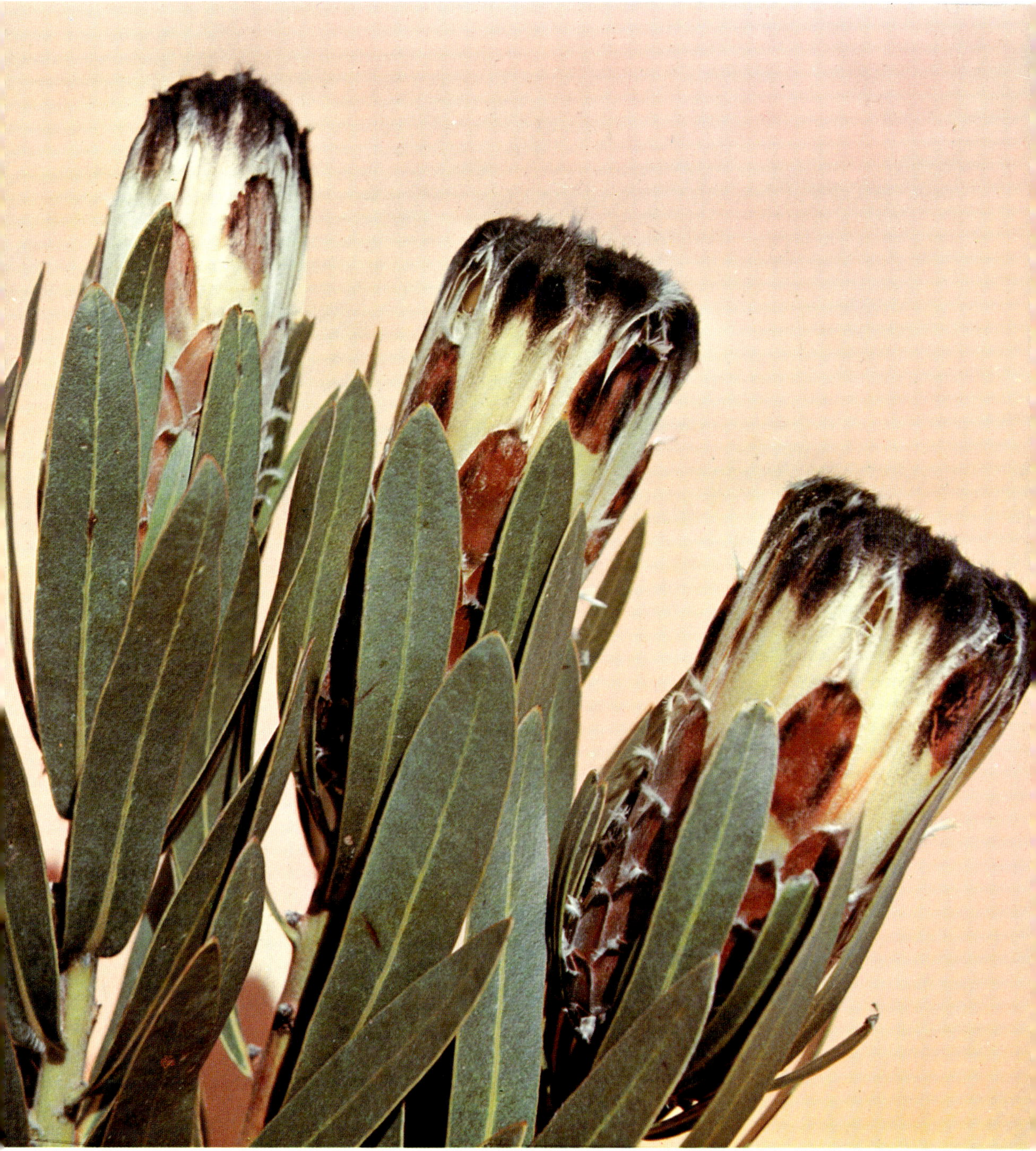

SUSAN'S PROTEA
Protea susannae

It is so easy to condemn a plant by accentuating an unpleasant feature and to discourage its cultivation, that one hesitates to mention it. The fact that the leaves of *P. susannae* have an unpleasant smell when cut or placed in a vase, however, should not blind one to the attractive features of the flower. One could strip the foliage before arranging the flowers if the odour is too noticeable, and this is done by discerning florists who find it a valuable cut flower and such a prolific bloomer. It flowers freely in autumn, winter and spring.

P. susannae has a very pretty bud, measuring about four inches long, in rosy-pink shading to brown at the base. This brownish tint is noticeable in the open flower-head too and gives it a pleasant old-rose quality. The flower-head is at its best when it is three-quarters open, for once it is wide open, the central mass of hairy pinky-white flowers fall outwards in rather an untidy fashion. The five-inch-long green leaves half enclose the head and should be removed to display this Protea in the vase. The bush is fairly compact, growing up to 6 feet in height, and may be pruned back to keep it in shape.

One of the most interesting features about the cultivation of *Protea susannae* is the fact that it grows in limestone soil in nature, near Bredasdorp in the south-western Cape. This enables gardeners who live in places where the soil is alkaline to have the pleasure of growing a Protea, even though it is not one of the most desirable species. This Protea will also grow in ordinary garden soil that is slightly acid and seems to present no problems in cultivation. Drainage should be reasonably good and the plants should be watered regularly, particularly during winter and spring.

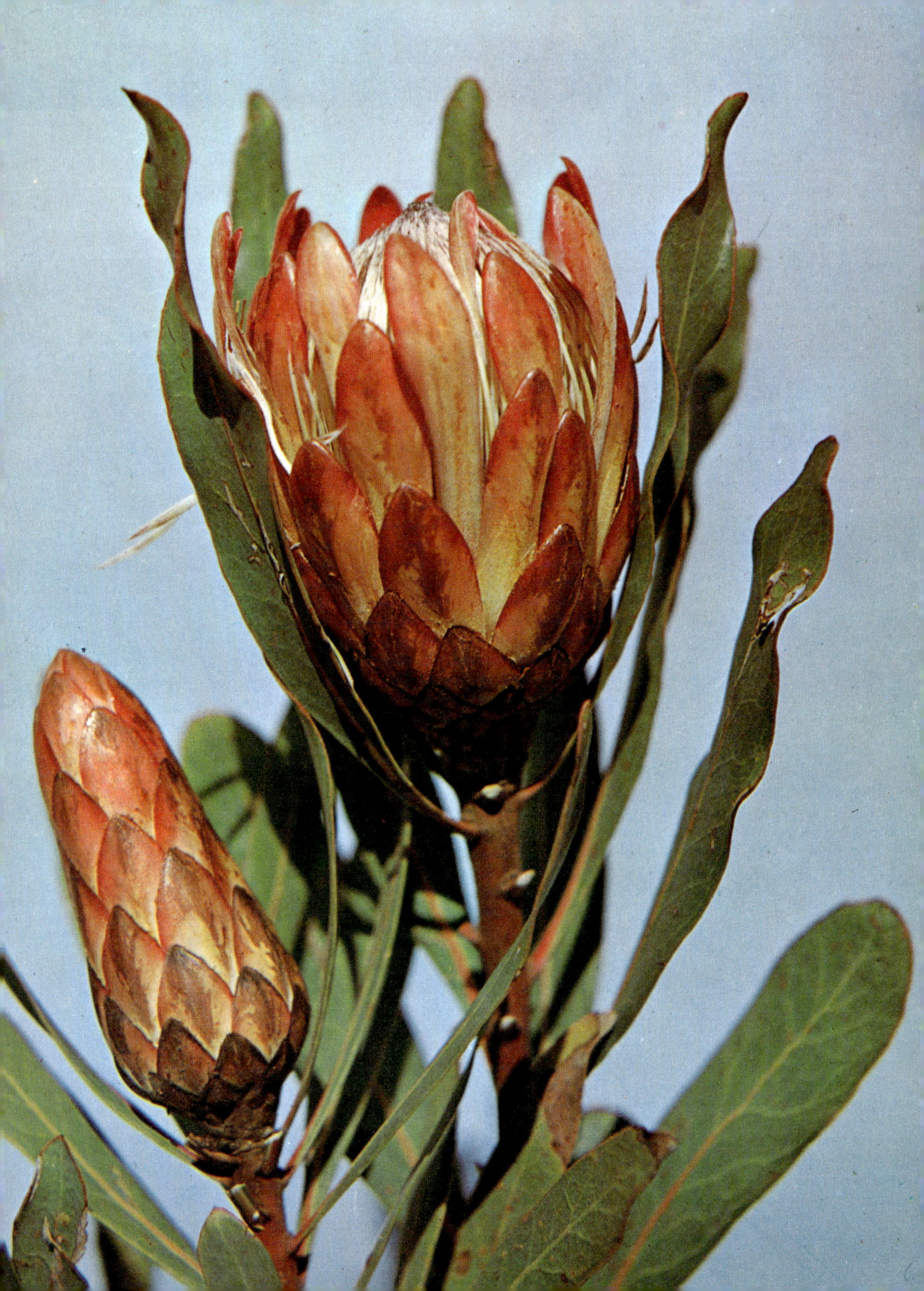

RAY-FLOWERED PROTEA

Protea eximia (*P. latifolia*)

This is one of the large-flowered Proteas that is grown quite commonly in gardens. When the bloom is fresh it can be beautiful, whether one has the rich reddish-rose or pinker colour variety. It becomes untidy as it ages. As in many types of Proteas, the colours vary to very pale shades, but as the colour is one of the most important horticultural features in *P. eximia*, one always hopes to obtain one of the stronger shades. The bracts themselves are spoon-shaped, swelling to just over half an inch across near the top and narrowing at the base. They are pink at the top and often shade to pale green near the base. Although they have no beard at the tips, each bract is softly outlined with fine white hairs. The centre of the flowerhead is thickly massed with pink hairy flowers that are darker at the tips, forming a broad spot. The head itself measures about five inches across at the widest part.

P. eximia is at its best in September and October, but blooms appear at intervals almost throughout the year, as this species has a long flowering period.

It is a great pity that the name *latifolia* must be discarded in favour of *eximia*, especially as the latter name means that it is distinguished for its size or beauty, for *latifolia* describes the broad leaves so well and makes the plant recognizable even when it is not in bloom. The silvery-green leaves are broad and oval, without stalks, and thickly wrapped around the upright branches. They grow up to about $4\frac{1}{2}$ inches in length, and over 1 inch in width. The whole bush grows from 6 to 8 feet and is fairly dense in appearance.

P. eximia grows easily and may flower in the second year if conditions are favourable. It will tolerate a large variety of soils, but does better in acid than in alkaline soil, needing reasonable drainage. It should be watered regularly, especially during winter and spring. It is quite cold-resistant. This species grows wild in the south-western Cape mountains, being distributed over a fairly large area.

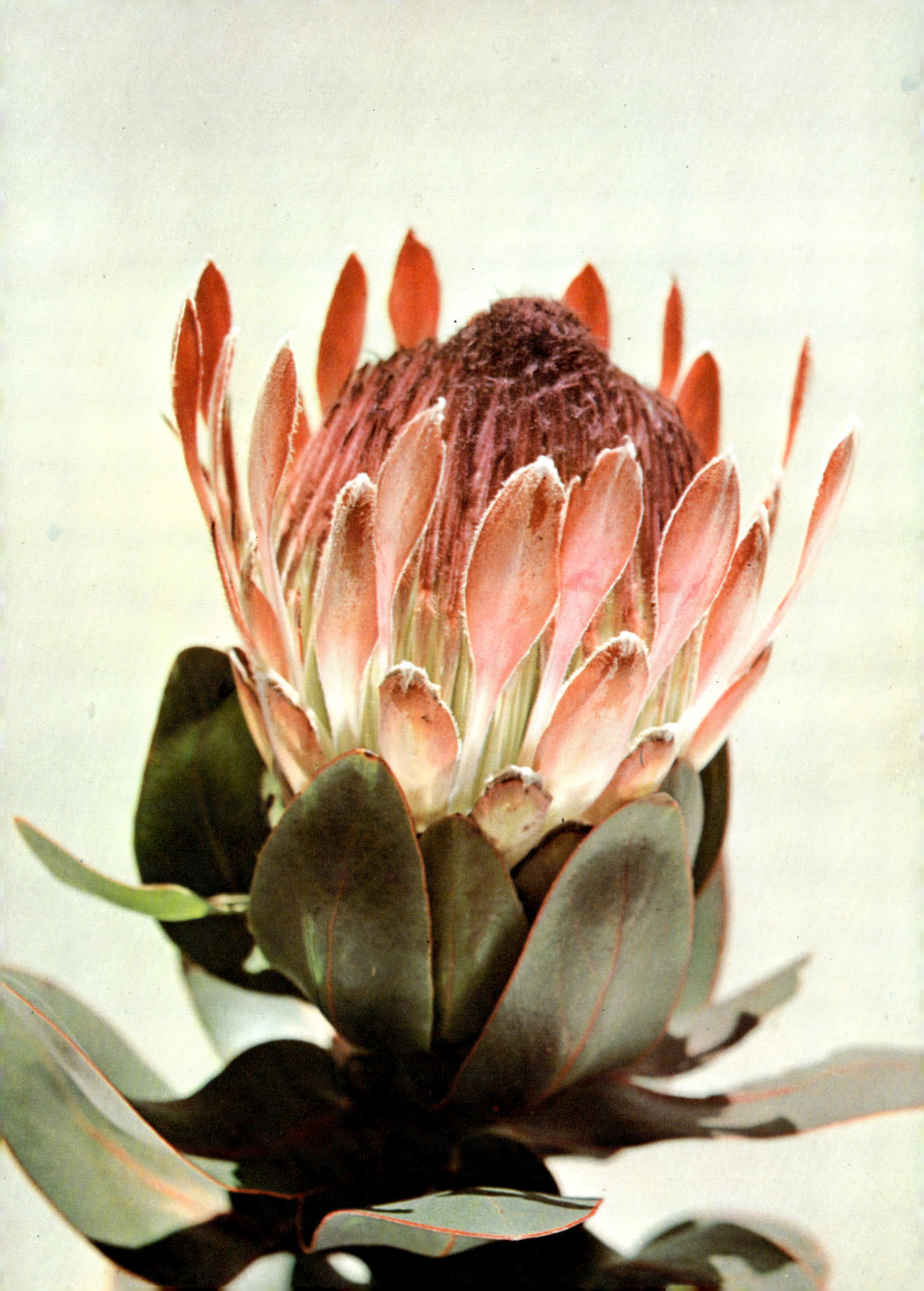

STOKOE'S PROTEA
Protea stokoei

One of the oustandingly beautiful Proteas, this has bright pink bracts tipped with dark brown fur. It measures about six inches in length. It is very similar to *P. speciosa* except that the silky beards are much shorter and there are small botanical differences in the inner flowers. The leaves too, are broader and more oval in shape. The bush grows to a height of about six feet.

This winter and spring-flowering Protea is not yet available in the trade, but would be well worth cultivating if it were. It is rare in nature and found in a small area at high altitudes in the southern Hottentots Holland Mountains in the south-western Cape. It is named after the late T. P. Stokoe, a mountaineer and collector, who found many interesting plants when botanising in the mountains of the south-western Cape, still continuing his beloved hobby when he was over 90.

The area in which these plants are found is cool and constantly bathed in cloud mists, even during summer. It would be necessary to give this species a cool position, with plenty of moisture at the root, yet protect it from frost in areas with dry, cold winters. The soil should be well-drained and contain compost. The flower shown in the photograph was grown at Kirstenbosch.

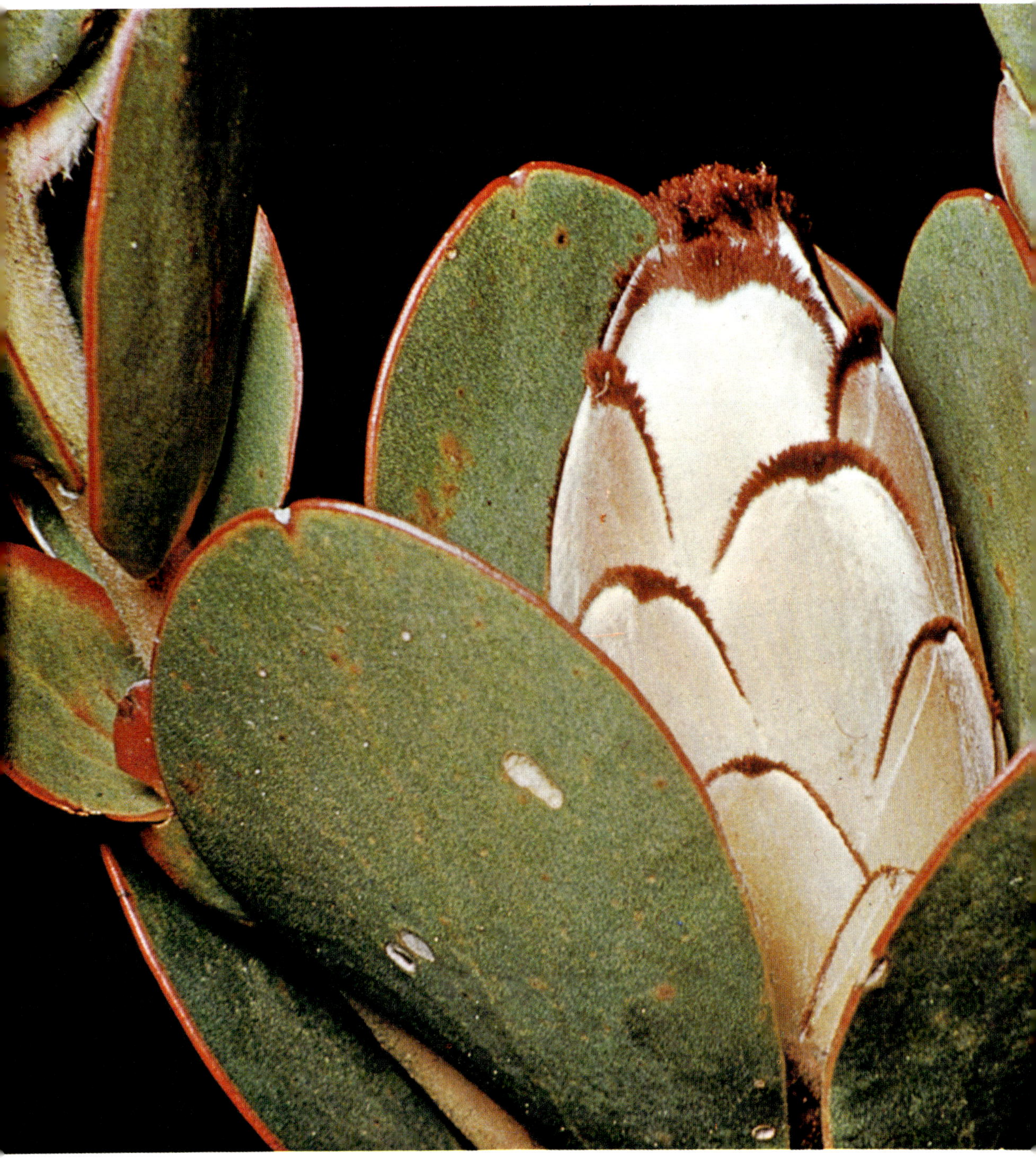

GLEAMING PROTEA

Protea burchellii (*P. pulchra, P. subpulchella, P. pulchella*)

A plant that starts blooming when it is only 2 years old and bears dozens of flowers on a low bush of only eighteen inches, is worth consideration by any gardener. When these blooms are exquisitely formed and have a metallic sheen, they are even more desirable. They last for up to 3 weeks in water and are best displayed in the vase when the leaves are stripped from around the buds and flower-heads. They bloom in early spring.

The flowers of *P. burchellii* vary in size. The bud is pointed and grows from 4 to 5 inches in length, opening into a fairly narrow cup shape. It is sometimes difficult to distinguish between the flowers of *P. burchellii* and those of *P. repens*, but the former has bracts with more rounded tips and each bract is tipped with dark fur, whereas the bracts of *P. repens* are completely smooth and pointed. The colours vary from a pale salmon pink to a deep ruby, with many subtle shades between, as well as green, cream and white. Strong colours are generally favoured in the garden, but the lighter colours show up better in the vase. The forms come true from seed. The leaves are about 4 inches long and half an inch in width, covered with tiny hairs, unlike the smooth narrower leaves of *P. repens*. The botanist recognises that the leaves have thickened margins, but that they are variable.

One of the advantages of *P. burchellii* is that it forms a low spreading bush that is particularly suitable for a rockery or any mixed bed in the garden. It will grow up to 6 or 7 feet in favourable conditions, but can be kept low by constant pruning, which will be necessary when cutting off the dead flowers. It grows extremely easily and survives severe cold, falling to zero in winter, if it is given a sunny position. It will grow in a mixed bed of flowers, which are watered frequently, without ill-effect, provided that the soil is reasonably well-drained.

This species grows wild on the lower slopes of the mountains of the south-western Cape. It occurred on the Cape Peninsula in former times, but has since been exterminated due to the spread of civilisation.

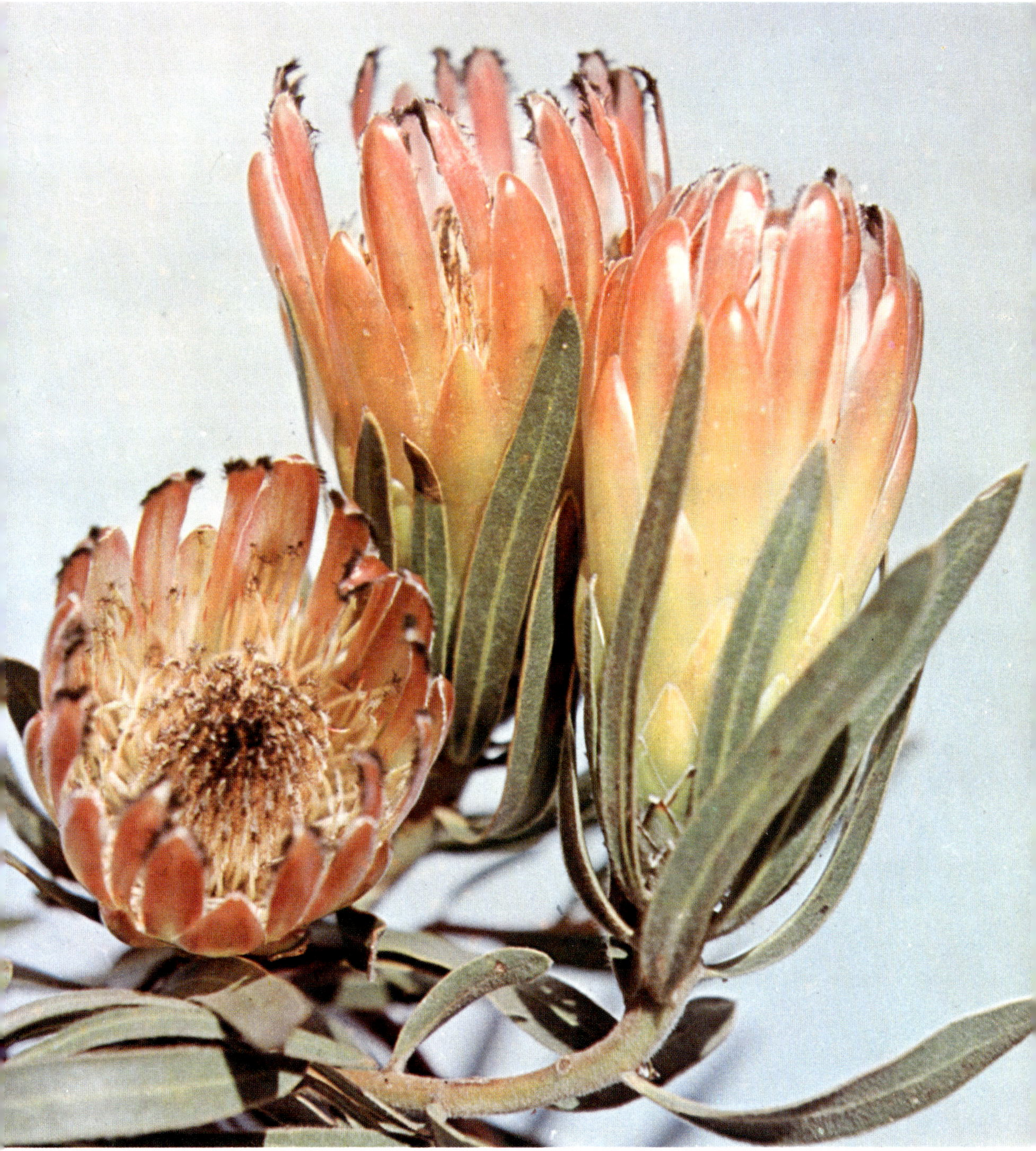

BABY PROTEA
Protea lacticolor

A delightful small flower for the vase, this grows on a large rounded bush that is suited to the larger garden. The charm of the wide-open, starry pink flowers has made it very popular with florists in recent years. They should be picked in bud and will open quickly in the vase, lasting for a week in water. The slim pink bud is only $2\frac{1}{2}$ inches long. When the flower is open, the delicate pink bracts resemble the wide-spread petals of a daisy, revealing a starry mass of pinkish-cream hairs within. *Lacticolor* means milky, but the flowers are really a delicate pink.

P. lacticolor flowers during spring and throughout the summer until March and April. It is covered with hundreds of buds during flowering time. The bush is large and rounded, growing up to 8 or even 10 feet in height. The leathery green leaves are smooth and undistinguished, growing up to about 2 inches in length. Cultivation is very easy and it is drought-resistant when mature. It needs sunshine, ordinary well-drained soil and regular watering while young.

This species grows wild in the eastern part of the southern Cape, on the Amatole Mountains near Hogsback and the Katberg mountains, as well as near Stutterheim and Keimanshoek. Natural hybrids form easily, especially with *P. mundii*, with which it should be compared.

P. punctata is almost identical to *P. lacticolor*, but has deeper pink flowers and smaller leaves. It is very difficult to distinguish between these two species except for the flower colouring, which is not a reliable botanical feature. *P. punctata* flowers in April and comes from the Cedarberg and Swartberg mountains, as well as from Aasvoëlberg near Willowmore, Cape.

P. subvestita is also very like *P. lacticolor*, but may be recognised by its silvery flowers that are edged with white hair, while the young leaves are hairy. This summer-rainfall species grows wild on the Drakensberg and flowers during summer.

P. parvula has similar flowers to those of *P. lacticolor*, but they are whitish and smaller. This is a low, prostrate species that would be suitable for rockeries in the summer-rainfall area if plants were obtainable. It grows wild on the eastern Transvaal escarpment, near Lydenburg and Pilgrim's Rest.

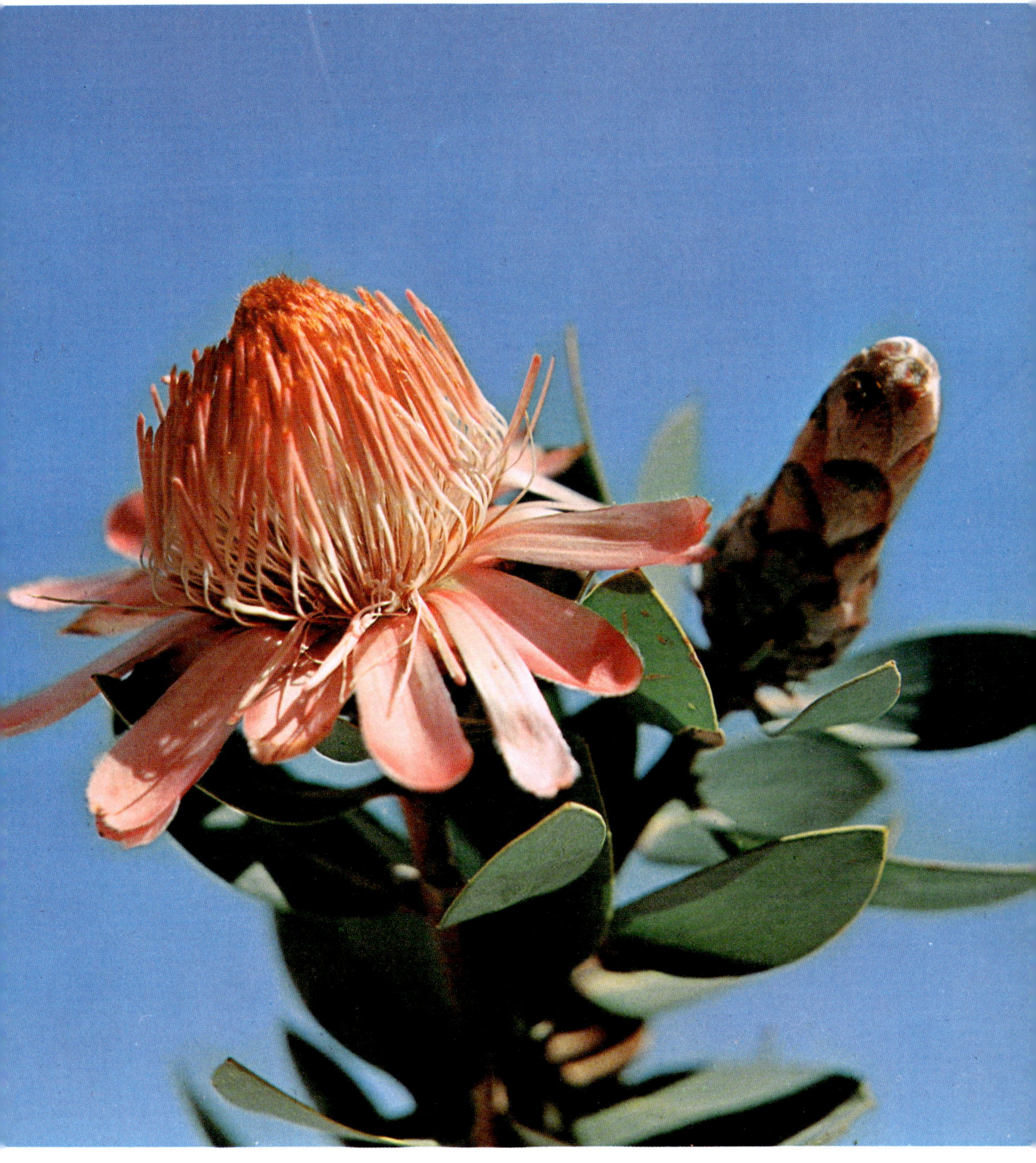

MUND'S PROTEA

Protea mundii

This Protea is similar to *P. lacticolor* in many botanical features, but the layman can generally distinguish between the two by the appearance of the flowers. Whereas *P. lacticolor* opens wide like a daisy and has charming baby pink colouring, *P. mundii* does not open quite so wide. It is not nearly as attractive from a florist's or gardener's point of view. Many natural hybrids exist and the true *P. mundii* is difficult to pin-point, but the flower illustrated here shows its most attractive form. Some of the hybrids have a very stiff appearance.

The flower measures about $2\frac{1}{2}$ inches in length, roughly the same length as the bud of *P. lacticolor*. The botanist recognises little knobs on the styles that are not present in *P. lacticolor*. To the layman, however, the bracts appear different, being slightly more pointed and hairy. The flower is attractive when fresh but does not last long. It is generally pink or white, but there is a good red form. The flowers appear in autumn and spring.

This is a very big, rounded, quick-growing shrub that often starts blooming in the second year from seed. It may reach a height of 6 feet in 18 months and will grow to 10 feet or even double that size in areas of good rainfall. The branches should be shortened while the plant is young in order to keep it compact, otherwise lanky branches give it an untidy appearance. The deep green leaves are leathery and widely spaced along the sometimes reddish stems. This shrub grows easily in a variety of soils, including heavy clay, but is better with good well-drained soil that is slightly acid. It needs sunshine and regular watering.

P. mundii grows wild in the south-western Cape in the Knysna and Plettenberg Bay area, where it receives good rainfall throughout the year, but also grows in other parts of the Cape near Port Elizabeth, Uitenhage, Humansdorp and Prince Albert, as well as in kloofs near Hermanus.

P. lanceolata is a similar type of Protea that grows to 6 feet and is offered in the trade. It has rather untidy white flowers.

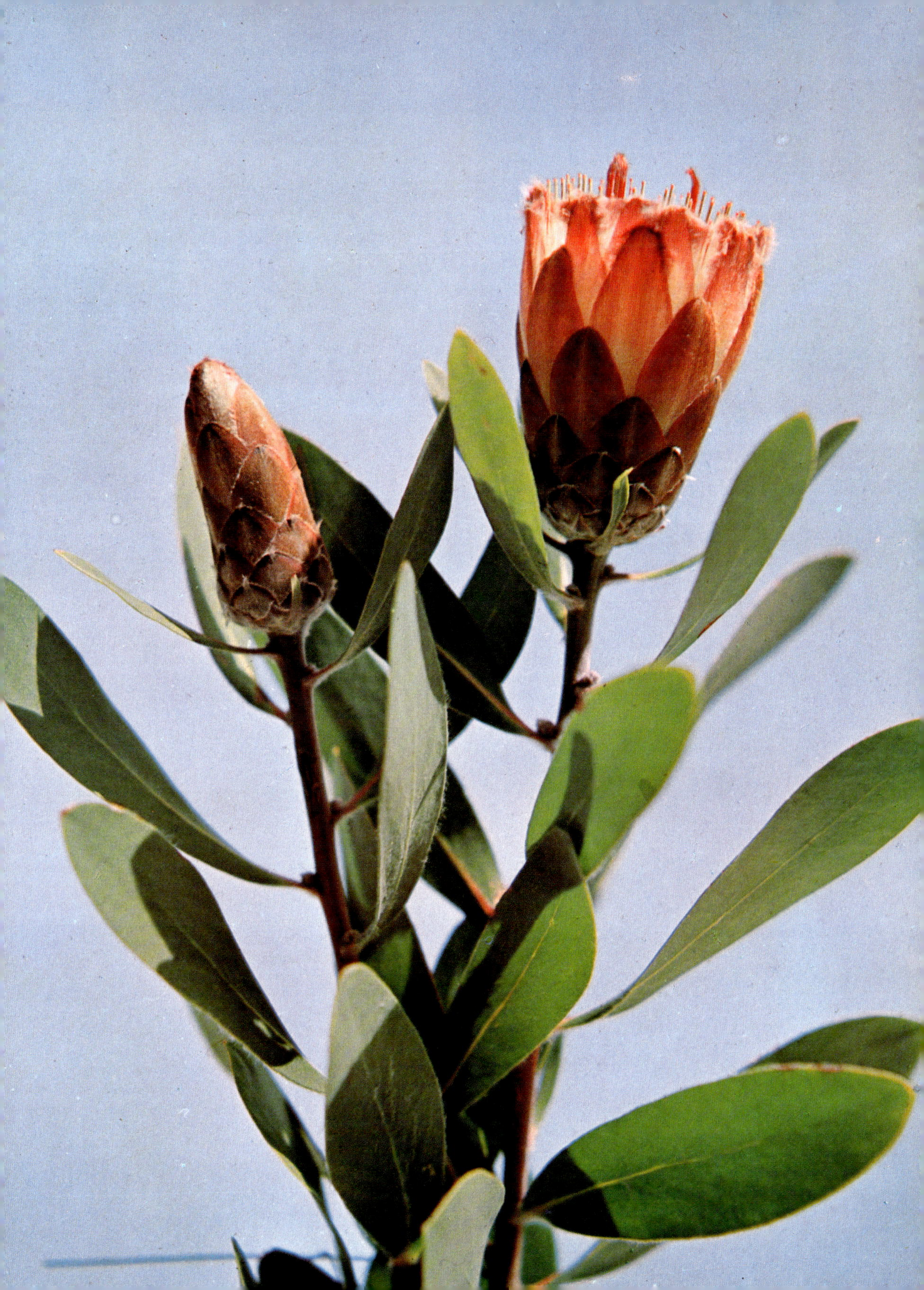

AARDROOS, GROUND-ROSE

Protea pudens (*P. minor, P. longiflora* var. *minor*)

Truly one of the most charming of all Proteas, this has a small flower-head measuring about three inches in length. Its daintiness and exquisite delicate colouring make it the focus of attention wherever it is exhibited. The rose-coloured bracts are slightly tinged with mauve and form a perfect cup. This is filled with a mass of pinkish-white hairs, tipped with brown or black, which rise to a gentle point in the centre. It flowers during autumn, winter and spring.

The leaves of this delightful flower enhance it by their own daintiness, for they are very long and thin, less than a quarter-inch wide and up to six inches long. They hang downwards on one side of the stem in the vase, but point up on the bush.

This is a low spreading plant with drooping branches that should not be propped up or staked in the garden. It should rather be planted in a rockery where it can droop down over the rocks or be allowed to spread to its full extent of three to five feet on a sloping bank or at the top of a terrace. One can pinch out the growing tips to encourage branching.

P. pudens needs a well-drained soil and grows in alkaline soil in nature, but will tolerate ordinary, slightly acid garden soil. It needs a sunny position with protection from frost in cold areas. Do not allow neighbouring plants to encroach on it or it may die. It grows slowly at first, but flowers while young. It grows wild in the Bredasdorp and Caledon areas of the south-western Cape. *P. pudens* is by no means the smallest flowered species. This distinction belongs to *P. odorata*.

P. odorata has a tiny pink flower, not more than $1\frac{1}{2}$ inches across, which flowers in spring and has a delicate perfume. It grows on a small compact bush about two feet in height. The leaves are stiff and needle-like, about an inch long.

P. odorata grows wild in a very restricted area between Malmesbury and Milnerton, near Cape Town. It has not been seen growing wild for many years and was thought to have become extinct until in 1954, when, through the efforts of several co-workers, it was rediscovered. Plants are now growing at Kirstenbosch and this species, no doubt, will find its way into cultivation.

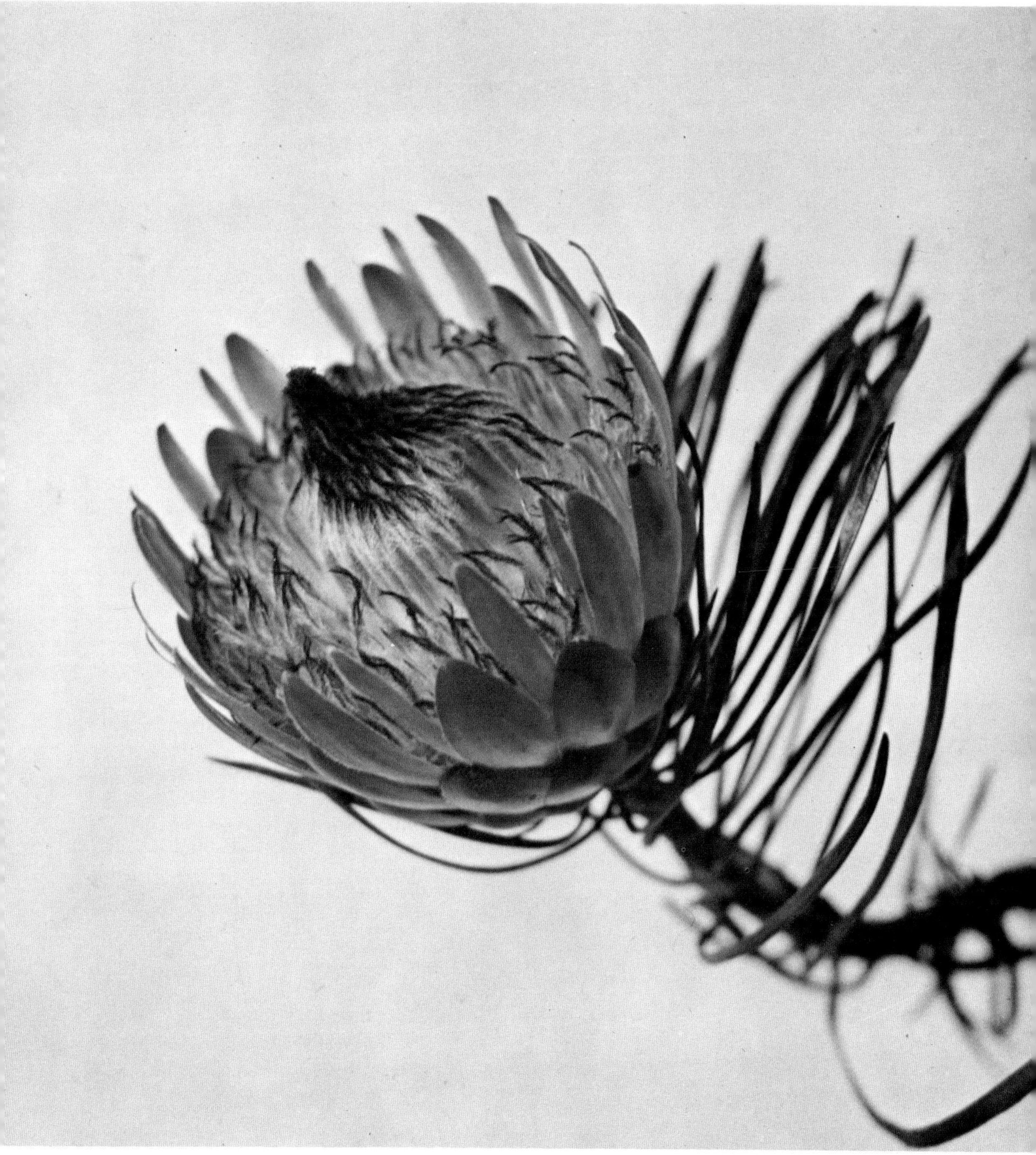

MOUNTAIN ROSE, BERGROOS
SKAAMBLOM
Protea nana (*P. rosacea*)

A charming neat shrub that could be grown in any garden, *Protea nana* is covered with nodding wine-crimson flowers in springtime. Each small bowl-shaped flower droops shyly from the tip of the branch and is seen to best advantage when the sun shines from behind to light it. The flower measures from 2 to $2\frac{1}{2}$ inches across and has overlapping ruby-coloured scales, often tinged green, that act like a tiled roof to keep off the rain. Its common name of Skaamblom or shy flower is taken from the bowed appearance of the little head.

The foliage makes this shrub easily distinguishable from another similar species, *P. pityphylla*. The soft needle-shaped leaves grow up to about an inch in length and cluster thickly up the stems, which bend gracefully like plumes in the wind.

The shrub begins to flower while quite small and a group of small plants of $1\frac{1}{2}$ to 2 feet is most attractive. When fully grown to 3 or $3\frac{1}{2}$ feet, *P. nana* becomes dense and compact, hung with scores of flowers which are most prolific in the spring. It flowers for a long period from June to November.

This plant grows easily in the garden, but needs very well-drained soil which is on the sandy side. It likes water during winter and will stand up to summer rainfall if drainage is satisfactory. Some protection from frost should be given in areas with cold winters. It is an excellent subject for a large tub, with its evergreen neat habit and dainty flowers.

P. nana is plentiful in the Cape mountains near Paarl, Worcester, Tulbagh and Ceres, where it grows on steep slopes. It is now freely obtainable from nurseries.

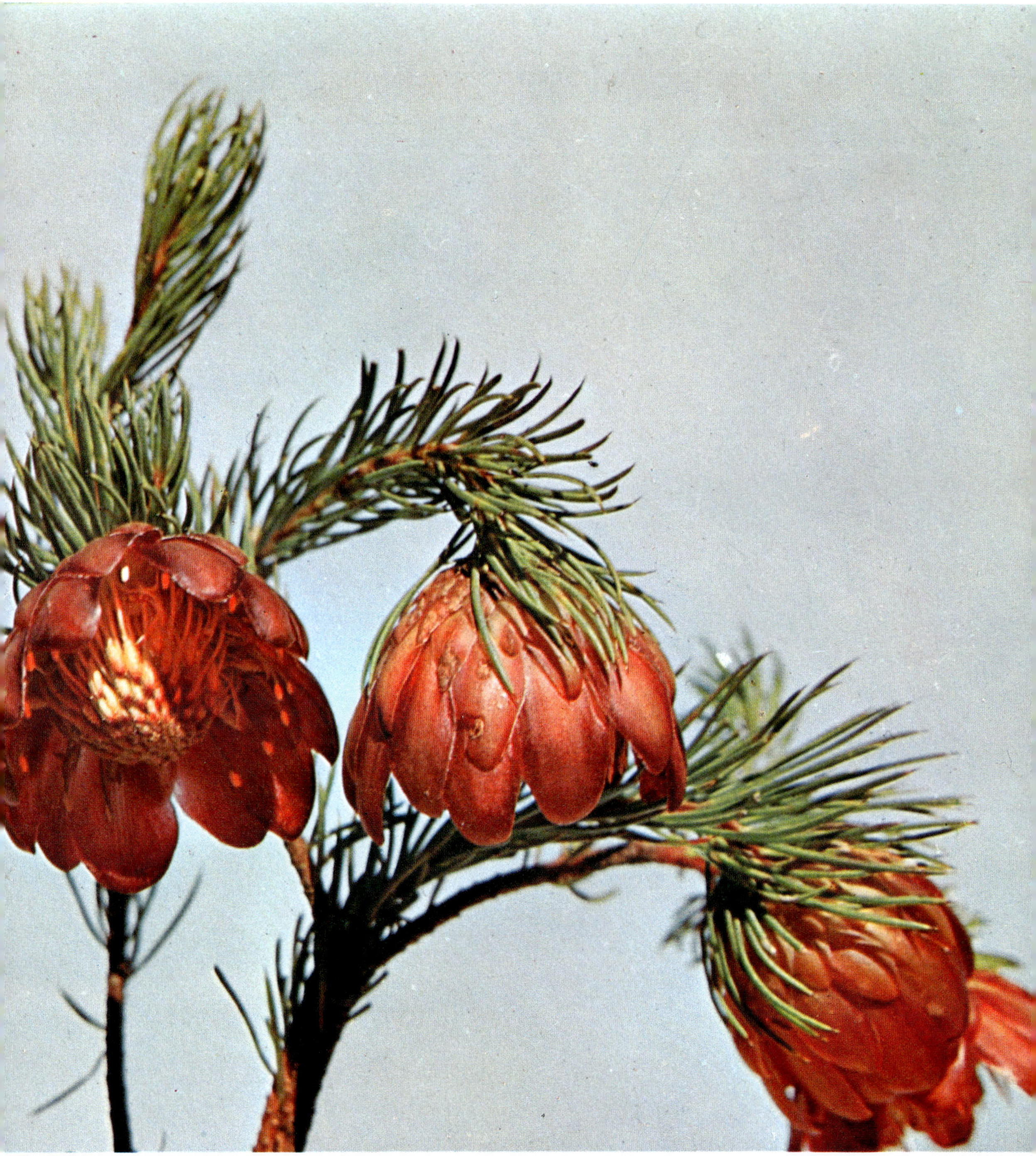

APPLE GREEN PROTEA

Protea coronata (*P. macrocephala, P. incompta*)

When they are well-grown, these apple-green flowers have a special beauty that is particularly valued by florists. The oblong head measures about four inches in length and two inches across and does not open wider. The outer bracts are a clear fresh green and curve inwards to rest on the mass of snow-white flowers that pack the centre. These come to a slight peak and the snowy tuft peeping out at the top enhances the perfection of the bloom. The topmost bracts are also edged with fine white hairs. Soft hairs also surround the base of the flower-head and extend down the stem. The flowers appear from autumn, through winter to spring.

The tapering leaves of this species are a soft greyish-green so that the bush itself is not striking in the garden. It is chiefly of interest to those who value the colouring of the flowers. The bush grows to a height of eight or ten feet and is fairly upright in its habit.

P. coronata grows very easily in the average garden, needing no special attention. It sometimes flowers in the second year from seed. It is suitable for a large rockery where there is good drainage and sunshine. It should be given regular water especially during winter. This species grows wild in the south-western Cape, on Table Mountain, and at Kirstenbosch, extending eastwards to Humansdorp.

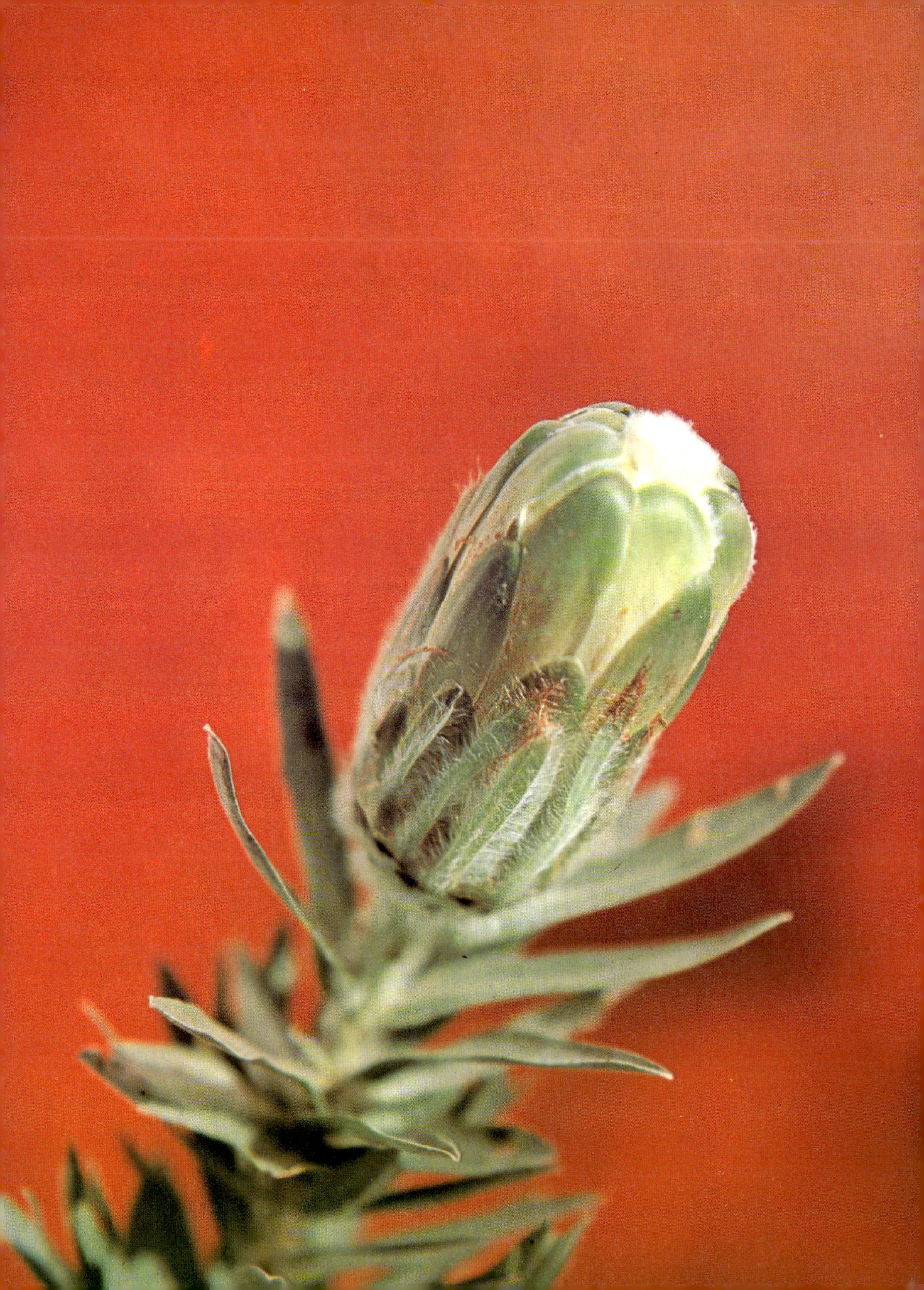

SMALL GREEN PROTEA
Protea scolymocephala

By no means a remarkable flower, this is valued chiefly by the flower arranger and the collector of interesting plants. It is a useful, compact evergreen shrub, growing to a height of about three feet, that should be given a place in a large garden.

The little flower-head opens wide to measure about $1\frac{1}{2}$ inches across and is coloured a delicate pale green. It has a certain charm when it is fresh and mixes well with the similar small flower-heads of the burgundy *P. acuminata*. The central flowers are not massed in *P. scolymocephala*, but look like a dainty puff of stamens. The flowers bloom prolifically for a long period and there are always round buds opening in autumn, winter, spring and early summer.

The narrow leaves are tough and leathery in texture, measuring up to two inches in length and less than a quarter-inch wide. They are densely massed on numerous short branches, giving the bush its compact density. Large branches sometimes bend and break from the weight of the foliage.

This is an easy species to grow, needing open sun, average well-drained soil and regular watering. It withstands drought and cold. It grows wild on hillsides on the Cape Peninsula, near Hout Bay and Camps Bay, as well as on the Cape Flats, and extends towards Malmesbury, Mamre and Somerset Strand.

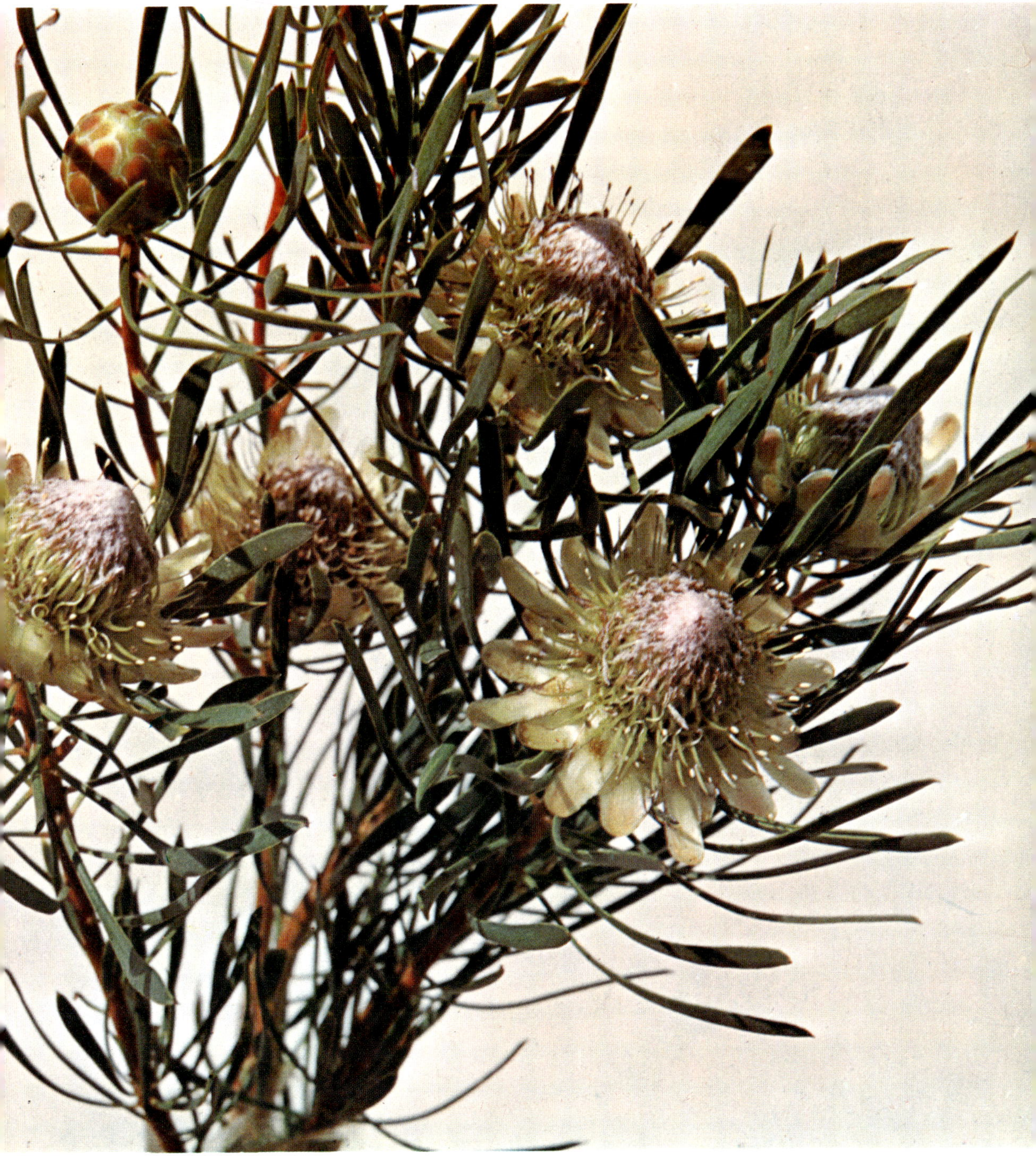

DWARF GREEN PROTEA
Protea acaulos

When stems of this green Protea were first seen at a florist, they seemed exquisite and it was hard to imagine that such a flower could be borne on an insignificant plant in nature or that it could be a botanical curiosity rather than have horticultural merit. It is an interesting plant for the collector who should grow it on a well-drained rockery. The flowers are attractive in themselves, having the shape of an old-fashioned rambler rose, and measure about two inches across when wide open. The curving bracts are coloured a delicate chartreuse and the central mass of flowers is pale green, tipped with old rose. One of the artistic features of this flower is the one-sided sweep of the leaves, which twist around the stem to face in one plane. This is caused by the leaves growing erect towards the light, but it adds to the value of the cut flower.

The plant has an underground rootstock and part of its main stem is also below soil level. The branches that appear above ground run along the surface like a creeper and the leaves point up at an angle. The flowers are borne at the tips or on the main stem and can generally be cut with a length of stem. The spoon-shaped leaves are about $5\frac{1}{2}$ inches long and taper from about two inches across to a quarter-inch near the base.

This species is grown at Kirstenbosch, where it may be seen flowering in the spring. Most of the species with underground stems are drought-resistant and should be given very well-drained soil in cultivation, especially if they are grown in areas where rain falls at a different season from their own. *P. acaulos* generally grows in dry sandy soil in nature. It spreads over a wide area from Cape Town to Clanwilliam in the north, and from the Stellenbosch area to Ceres, Swellendam and Bredasdorp in the east.

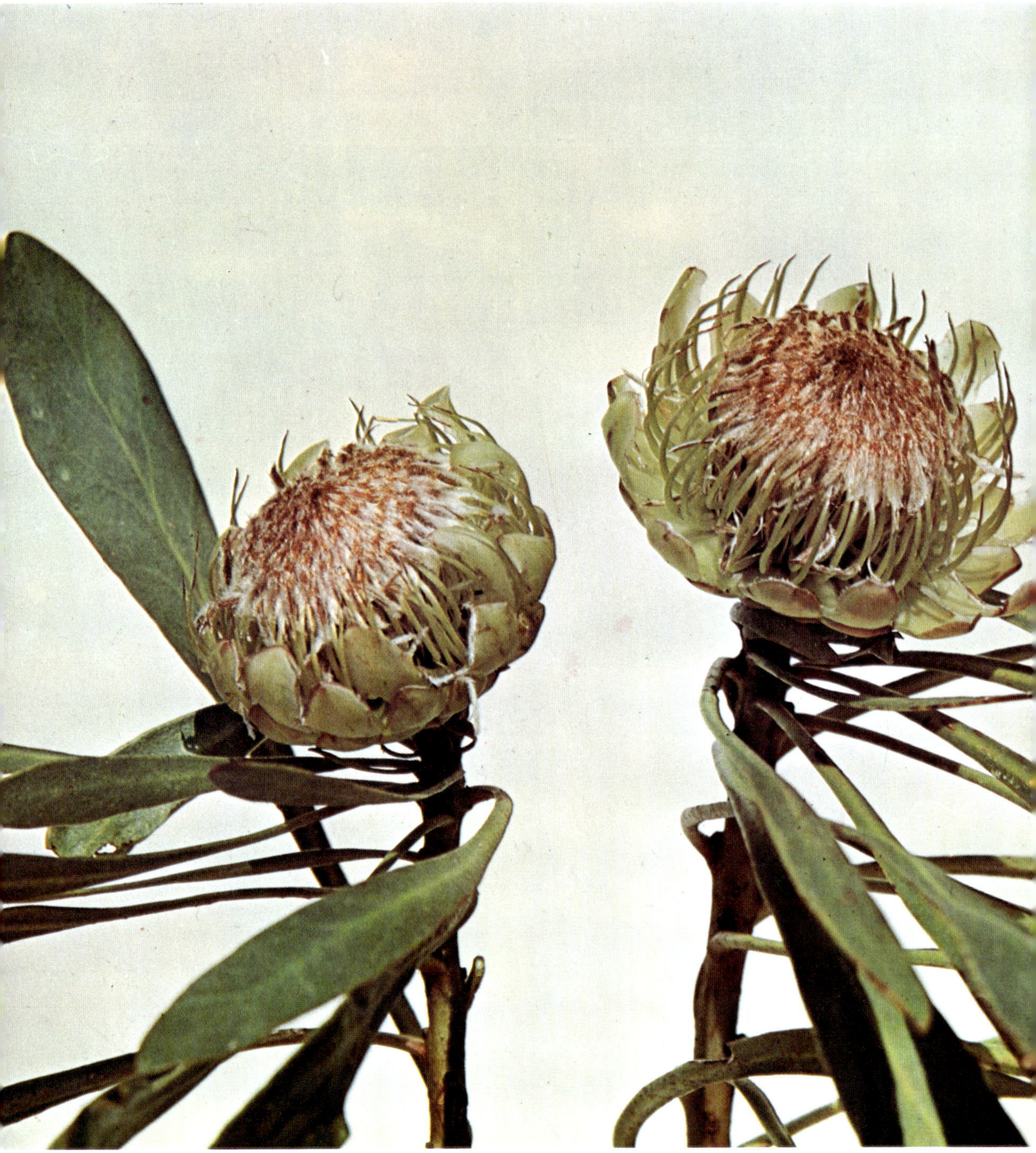

WITZENBERG PROTEA
Protea witzenbergiana

The bowl-shaped flower-head of *P. witzenbergiana* is very similar to that of *P. pendula*. It is a little smaller, measuring about two inches across, spreading wide to show the red colouring of the bracts inside. The outside of the bracts is smooth, unlike *P. pendula*, which is silky-hairy outside. The flowers form a densely packed beehive at the centre. This plant blooms in the spring.

This is a low, spreading shrub, from one to two feet in height, which bears its stiff flower-heads at the ends of the branches so that they face down towards the ground. The leaves stand up vertically towards the sky and look like gracefully curved needles. This is the chief point of distinction when it is compared to *P. pendula*. Comparisons have also been made between *P. witzenbergiana*, *P. nana* and *P. pityphylla*, but it is a spreading, stockier, bush, while the two Mountain Roses are upright in growth.

As the name applies, this species grows wild in the Witzenberg range near Tulbagh in the south-western Cape, on high peaks where snow falls in winter. It is obtainable in the trade. Cultivation does not seem to present any special problems, provided that the soil is well-drained and the plants receive water during winter.

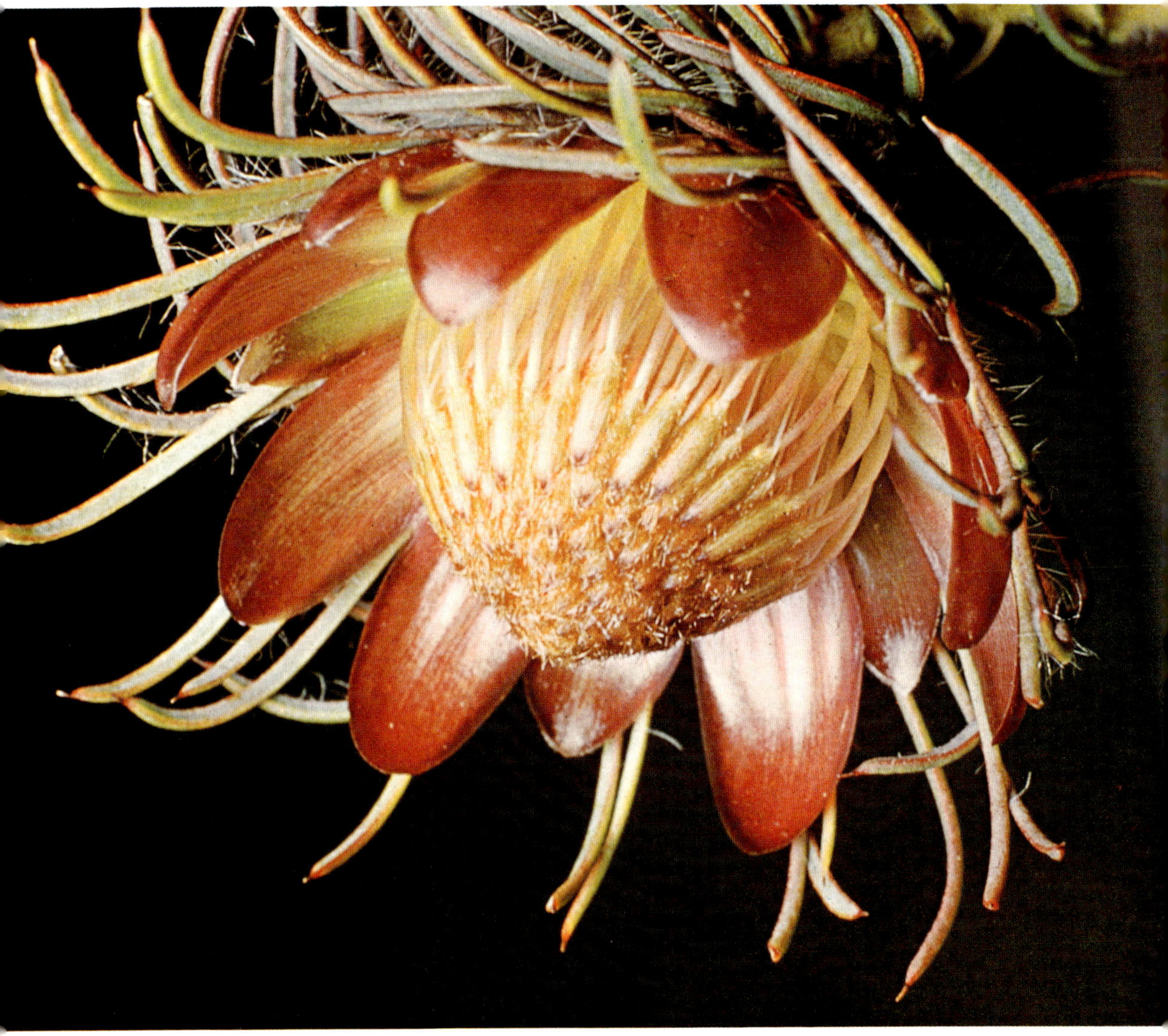

SWARTBERG PROTEA

Protea venusta

This is a spreading plant that would be excellent in a large rockery or on a sloping piece of ground. It is a fairly rare Protea that has been introduced into cultivation recently by one of our progressive nurserymen, who went to the trouble of collecting some seed high up in the Swartberg mountains of the south-western Cape, near the top of the pass. It also grows in the Kamanassie Mountains near Uniondale.

Snow falls here each winter, so that this species is certainly cold-resistant. It would, however, be advisable to shelter it from the dry frosts of the highveld by planting it within the shelter of other shrubs, but one must be careful not to shade it. As it grows in well-drained rocky places in nature, it should be given well-drained soil in the garden. Cloud mists provide water in summer so that water should be given regularly in dry areas, but this is a fairly drought-resistant plant. *P. venusta* has been grown at Kirstenbosch, where the plants flower very seldom.

P. venusta has exquisite colouring and is well worth cultivating. The flower-head measures about $3\frac{1}{2}$ inches across when it is open. It has deep, rose-pink bracts that shade to white near the base. They are velvety on the outside, with soft white hairs on the edges. The plant has a thick underground root that sends up branching stems to a height of about 18 inches. These then send out long stems that grow up to about 15 feet and drape down across the rocks. Each bloom appears on new growth at the tips of the spreading branches, turning upwards instead of facing downwards, as is the case with many sprawling Proteas such as *P. amplexicaulis* and *P. sulphurea*. This gives it even greater horticultural value than these species.

The oblong leaves grow up to about $2\frac{1}{2}$ inches in length and $\frac{3}{4}$ inch wide, tapering to the base. They stand erect as do those of many of the creeping types of Proteas.

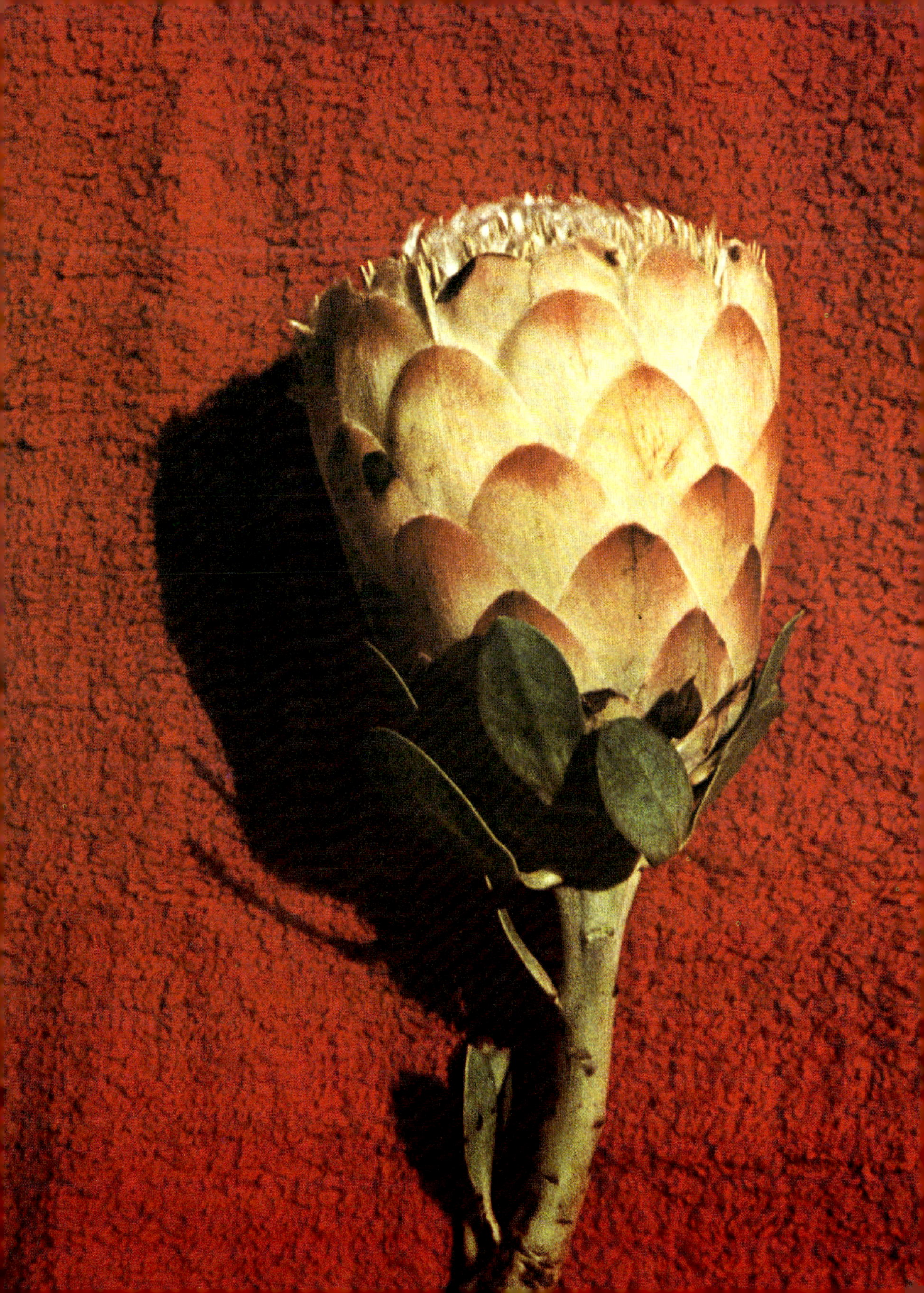

GROUND PROTEA

Protea humiflora

"Humi" means "on the ground" in Latin and this is one of the low-growing Proteas with an underground rootstock and main stem. In this case, however, the side branches do not creep low along the surface, but stand erect, sometimes reaching a height of almost two feet.

The flower-heads appear singly, being closely set against the woody stems, while some emerge at ground level. They measure about 2 inches across and the red hairy bracts curve inwards at first, hugging the dome of hairy flowers at the centre. These are yellow, tipped with red. When the bracts open wide, they are whitish near the base. The flowers bloom during winter. The leaves are curved and very thin, almost needle-shaped.

Like most of the Proteas with underground stems, this species grows in sandy soil in nature and would require very well-drained soil in a garden, being suited to conditions in a rockery. It is obviously a plant for the collector, but since it was cultivated before the year 1800 in England by H. Andrews, and illustrated in his *Botanist's Repository*, it is reasonable to suppose that modern enthusiasts would wish to grow it. It should certainly be included in any serious collection.

P. humiflora grows wild in the south-western Cape, high in the Swartberg Mountains and elsewhere, where it endures cold winters.

Protea scolopendriifolia (*P. scolopendrium*). This "ground" Protea has short stems that make it a plant for the collector rather than the horticulturist. The flowers are attractive when fresh, measuring about 8 cm across, with numerous, pointed deep rose bracts. The long narrow leaves, up to 20 cm in length, broaden near the top. This species occurs on the high mountains near Tulbagh and on other mountains of the S.W. Cape.

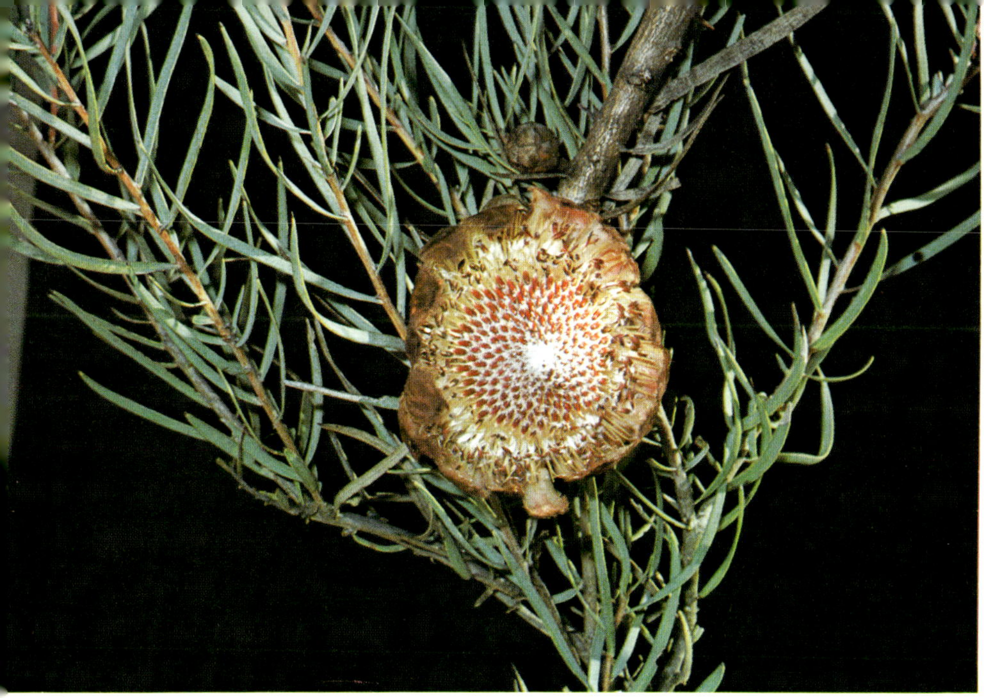

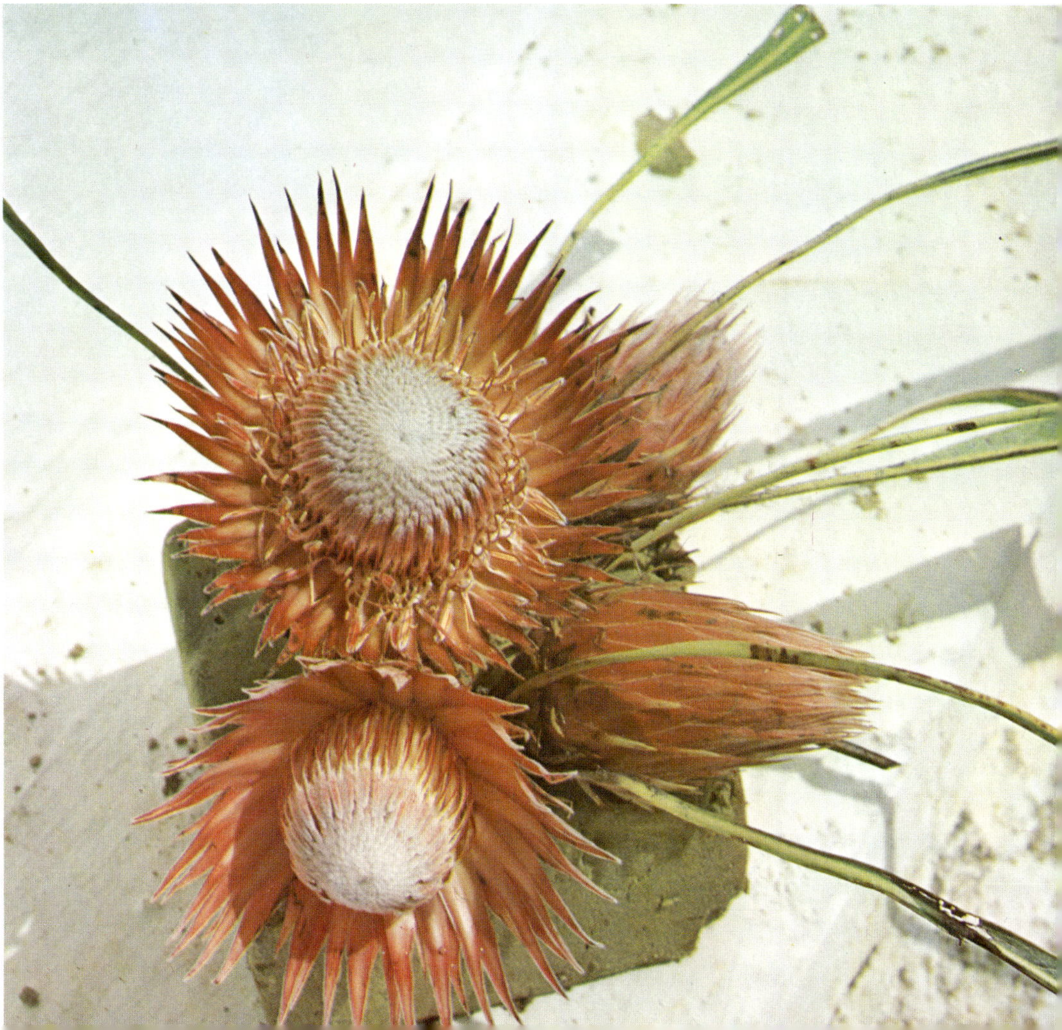

HEART-LEAF PROTEA
Protea cordata

A strange plant, this makes a rounded tuft of waving stems extending over an area of about three feet. The tiny flower-head itself is insignificant, for it appears at ground level and is stemless. The flowers sometimes lie under the surface of the sand in the Kleinmond area. The plant is valued for its attractive heart-shaped leaves that clasp the graceful red stems at spaced intervals. They are generally light green, veined with red, but the new growth is pink or even red. The leaf stalks last in floral arrangements for weeks.

This plant has an underground rootstock and main stem and the short branches that appear above ground die back each year at the end of summer. The flowers appear from midwinter to spring. They are small and reddish-brown, similar to those of *P. subulifolia*.

P. cordata would be most suitable in a well-drained rockery on a gently sloping bank. It needs very well-drained soil and grows in poor, sour, sandy soil in nature. It grows wild in the drier areas around Fransch Hoek and Houw Hoek near Bot River, in the Caledon and Bredasdorp Division, as well as in the mountains of Du Toit's Kloof and Bains Kloof at elevations up to 2,800 feet. It grows in sandstone in nature. This plant is not hardy to frost and needs a fairly sheltered position on the highveld. The leaves mark easily and must be protected from careless handling and from insects.

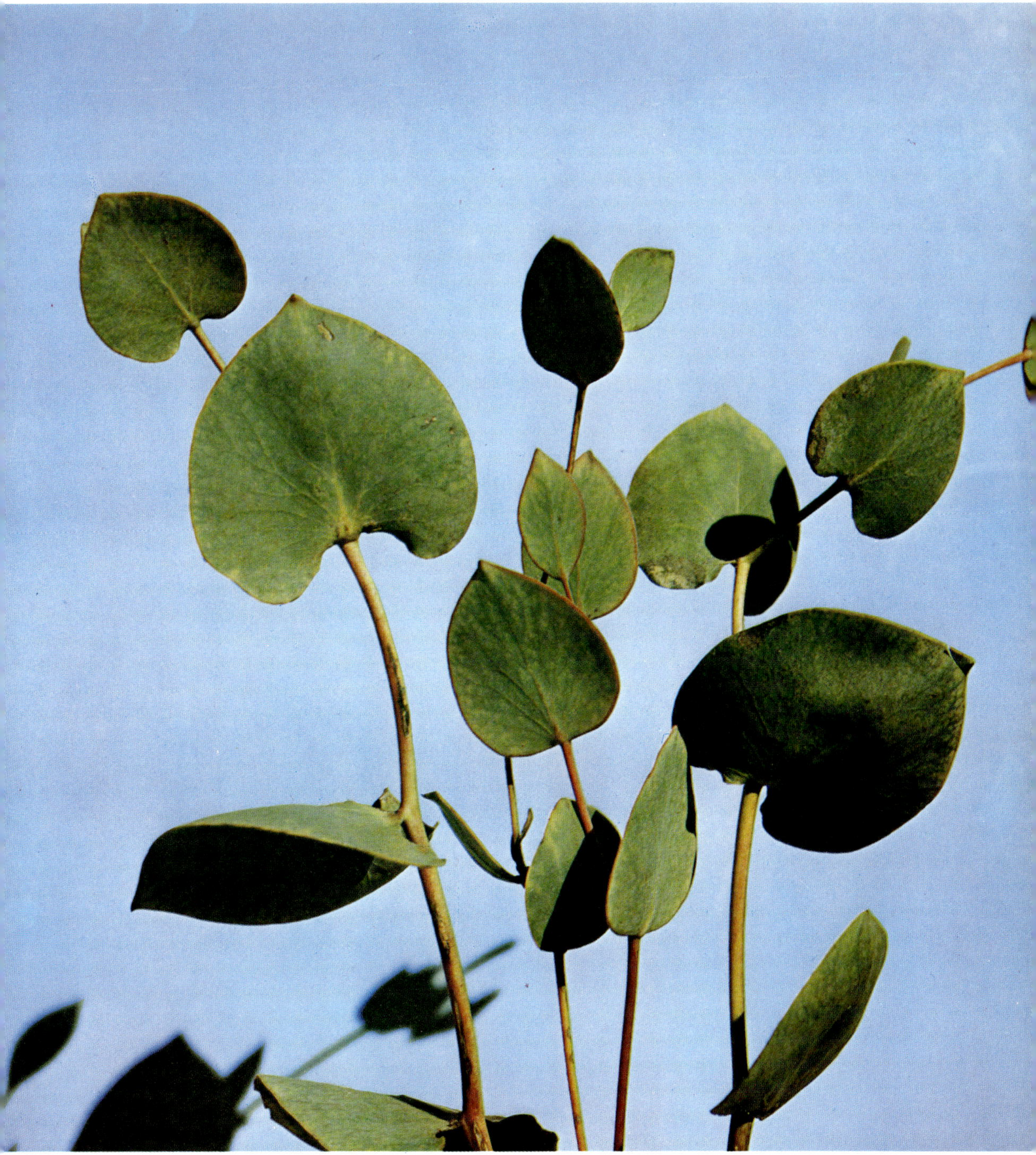

SNEEUBLOM, SNOW FLOWER, SNOW PROTEA

Protea cryophila

This giant snowy flower looks like a fluffy snowball itself and is called a Snow Flower because it grows amongst the snows in nature. It is found growing on high peaks of the Cedarberg Mountains, where snows fall every winter. The gardener in cold areas may feel that this is a good hardy species, but it is not likely that this plant will ever become popular in gardens.

It grows on a very large spreading bush which covers an area of about eight feet, mounding to a height of about two feet. The huge flowers emerge either at ground level or on a one-foot stem. Each flowerhead measures about eight inches across and the overlapping narrow bracts are covered in white wool. The central mass of white flowers seems to fill the head with white felt. The main stem is underground, but side branches emerge, bearing exceptionally long leaves that surround the flowers. They measure from 18 inches to 2 feet in length and are half folded or grooved down the centre.

Another Protea with long leaves is *P. lorifolia*, but these are not as large as those of the Snow Protea.

The Snow Flower is not the only Protea to grow at high altitudes. It has several others growing nearby in nature, including *P. acuminata*.

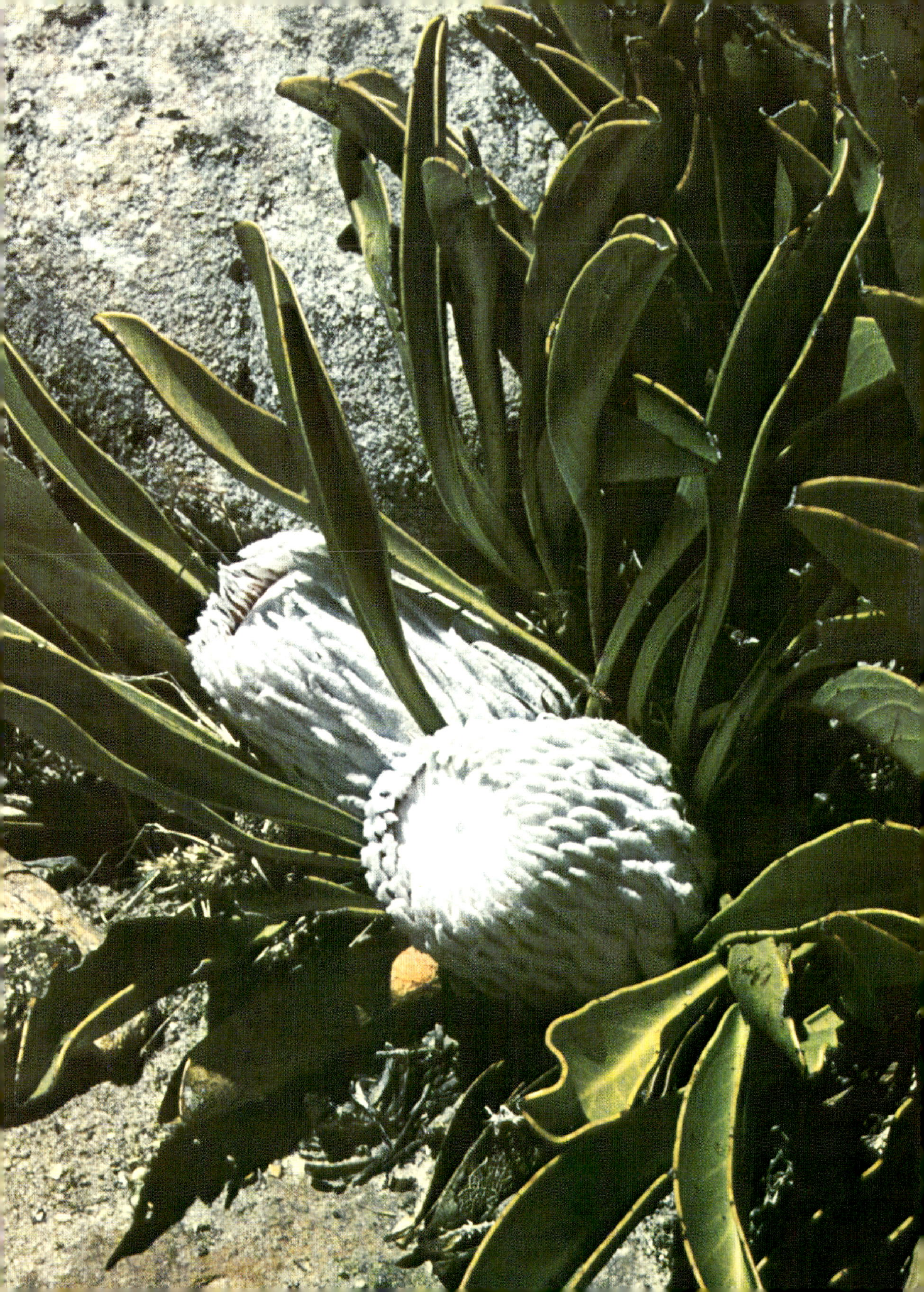

GRASS-LEAVED PROTEA
Protea restionifolia (*P. echinulata*)

This little ground Protea resembles a tuft of grass in the veld and it is not until the flower appears that one could recognise it as a Protea. The main stem lies underground, with the long leaves growing in tufts up to 6 or 8 inches long. The flowers emerge at ground level and are seldom seen more open than in the photograph. The brownish-pink flower-heads measure about 3 inches in length. An interesting plant for a rockery, this is a curiosity rather than a decorative garden plant. It grows wild in the Worcester and Caledon areas of the S.W. Cape.

There are several similar ground Proteas with small tufts of leaves, as distinct from those ground Proteas with spreading or rambling branches. These tufted types grow easily and would be suitable for smaller rockeries if seed were made available.

P. scabra. The flowers of this ground Protea are borne at the ends of the short branches, even though the leaves, which are grouped below the flower-head, stand well up above it. The browny-pink flowers are about 2 inches long and the pine-needle-shaped leaves are about 9 inches in length. Raised dots on the surface account for the name *scabra*, meaning "rough". This species grows wild in sandy soils near Caledon, Swellendam and Uniondale.

P. aspera is another ground Protea with long, narrow, grass-like leaves. These emerge at ground level with the reddish-brown felted flower-head nestling amongst them. The bracts of the flowers are edged with hairs. The leaves broaden slightly near the tip and so differ from those of *P. restionifolia*. This plant grows on sandy soil near the sea in the Caledon and Bredasdorp Divisions of the S.W. Cape. (Illustrated opposite, below).

P. lorea has the largest flower-head of these small ground Proteas. It grows up to 5 inches in length and seems incongruous emerging at ground level. The velvety bracts are pointed and a good pink colour. The flower-head opens more than those of the other ground Proteas, which seem to remain bud-like. The grassy leaves grow up to 9 inches in length. This comparatively rare Protea is found in acid soil at the foot of the mountains from Somerset West to Caledon and at Fransch Hoek.

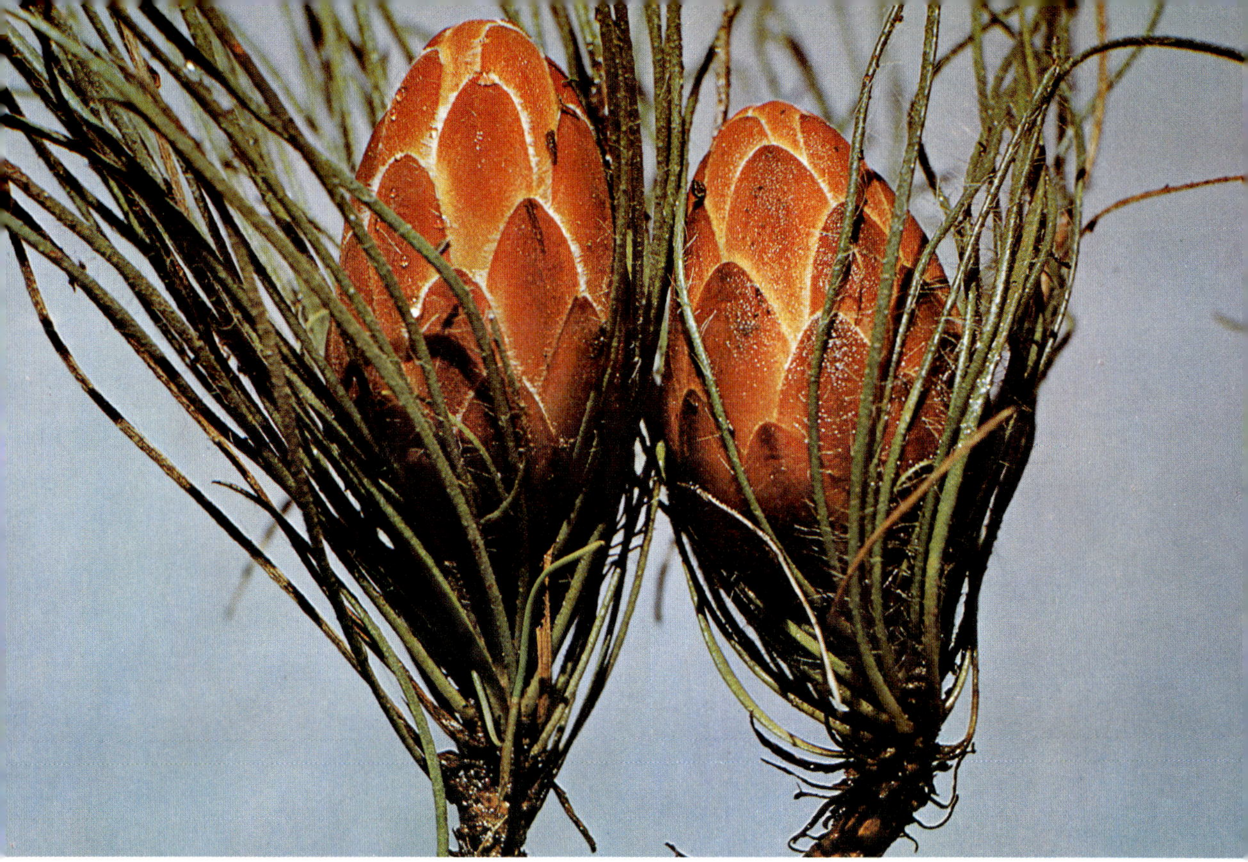

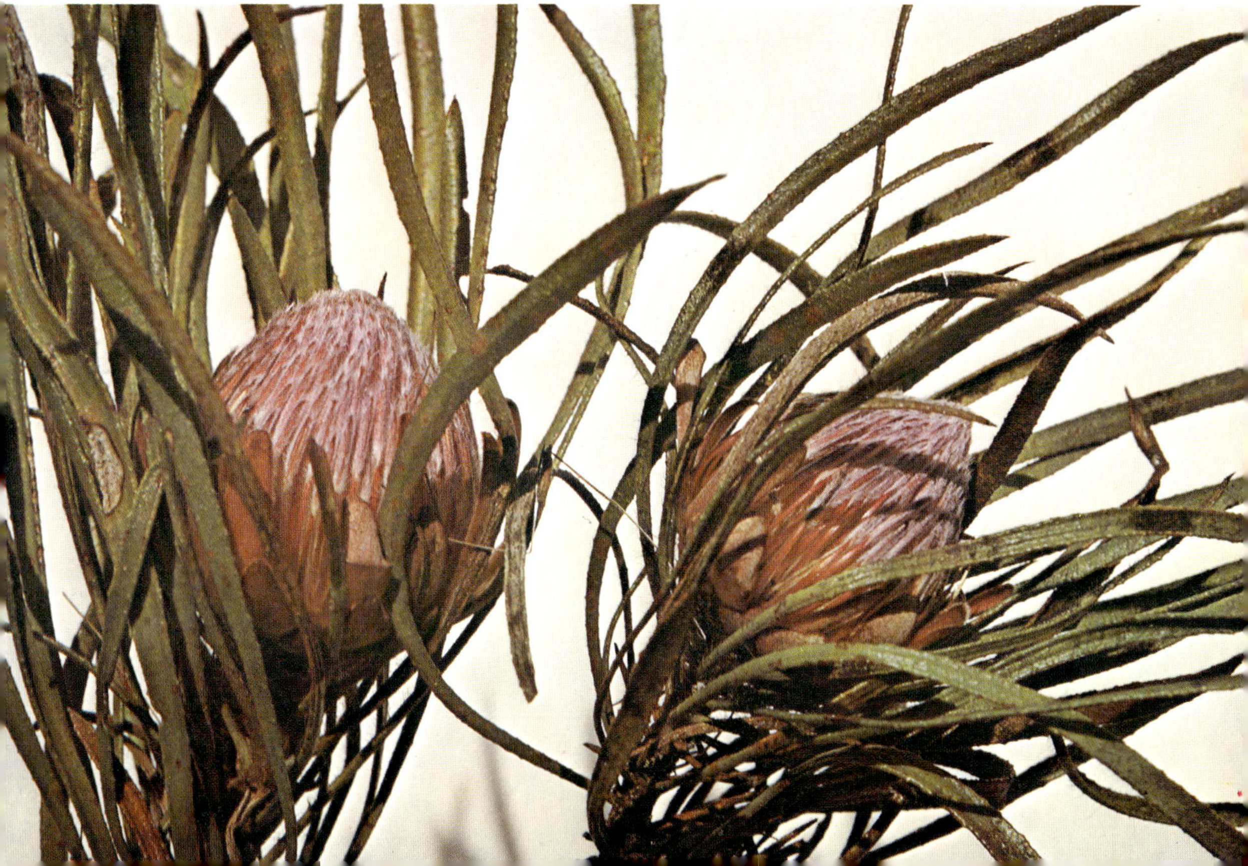

NODDING PINCUSHION
Leucospermum cordifolium (L. nutans)

Without doubt the finest landscape shrub in existence, the Nodding Pincushion not only makes an unequalled splash of colour in the garden, but provides scores of magnificent cut flowers that last for a full month in the vase without shrivelling.

It forms a neat spreading bush that grows up to about 4 or 5 feet and covers the ground over an area of about six feet. It nestles attractively among the rocks in a large rockery or will settle on the gentle slope of a bank with equal poise. Its size and neatness make it suitable for even a small garden, while several should be planted in a large garden. The flowers appear mainly in spring and early summer from September to November, but may start blooming in late autumn and winter in areas with mild weather. They appear later in areas with cold winters. The buds are sometimes frosted if the weather is severe, but new buds open later and bloom when the cold weather is over.

Unlike those of *L. reflexum*, the flower-heads do not change their shape as they mature. This makes them more valuable in that they remain decorative for a longer period both on the bush and in the vase. The dome-shaped flower-head measures about 4 inches across with the individual tube-shaped flowers curled like ribbons at the centre. From these, the glossy styles or "pins" spring out to form the attractive pincushion. The flowers may be pinkish yellow or orange and the styles vary in colour from delicate salmon to orange, yellow, apricot or flame, each tipped with yellow. A bright yellow colour form, formerly known as *L. bolusii*, is *L. cordifolium*, and *L. bolusii* is another species. The colours come true from seed, but it is not possible to know what colour you will obtain before the plants have flowered. As they are all beautiful, this does not seem to matter. They generally bloom in the third year from seed and bloom prolifically in the fourth, bearing a hundred flowers at a time.

The leaves of *L. cordifolium* are green and broad, measuring about 3 inches long and an inch in width. They are placed at regular intervals along the sturdy, gracefully curved branches. The bush remains attractive for about 9 or 10 years and flowers so abundantly that it generally dies after this period. It is wise to put in new plants of *L. cordifolium* every five years.

The Nodding Pincushion grows wild in the S.W. Cape in the Caledon, Bredasdorp and Swellendam Divisions. It is distinguished botanically from *L. vestitum* by its glabrous involucral bracts and "conic-ovoid" stigmas.

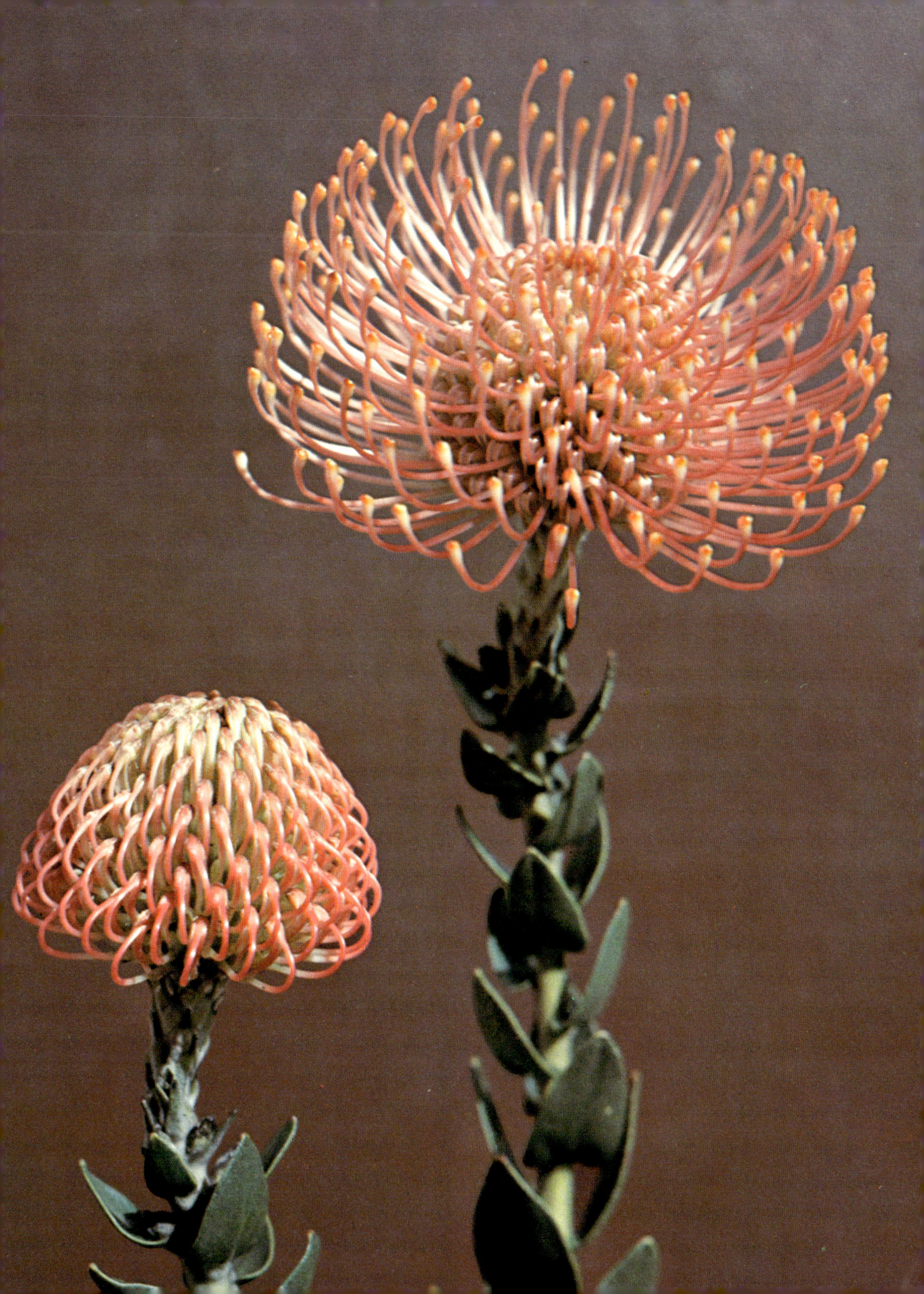

ROCKET PINCUSHION
Leucospermum reflexum

The majesty of a great grey bush of *L. reflexum*, covered with several hundred glorious blooms, is a sight never to be forgotten. With the background of the blue mountains at Stellenbosch, the breath-taking beauty of the shrubs and the scenery almost defies description.

But it is not only in the Cape that these magnificent shrubs may be seen. They grow robustly and luxuriantly in all parts of the country, thriving in the warm dry winters of the summer-rainfall area, provided that they are watered regularly during the winter months. They make a spectacular feature anywhere and look particularly attractive when grown in masses. The concerted effect of the orange-scarlet flowers, deepening to crimson as they age, and seeming to form rockets as they mature, is overwhelming.

The remarkable feature of the Rocket Pincushion is the way in which the flower reflex as they grow old. The buds unfold into a rounded pincushion of velvety tubular flowers. Each salmon-orange tube is yellowish at the base and covered with soft, silvery hairs, crowded especially near the pointed tip. A shining deep crimson style springs out after about a week to curl backwards and downwards. The flowers also curl outwards and resemble a pointed rocket-head with crimson streamers below it. The flower withers after about the second week and new flowers will open alongside it if it remains on the bush. Several buds at the tip may open if the flower has not been picked, or if the first flower has been damaged by frost.

The dove-grey foliage, consisting of small neat leaves pointing upwards all along the length of the stems, makes this a valuable landscape shrub. The leaves are about 2 inches long and $\frac{1}{3}$ of an inch wide. The flowers should be cut with long stems in order to keep the bush within bounds. It is most suitable for a large garden as it grows to a height of about 9-12 feet and needs at least 12 feet in which to spread, as it branches from ground level. It generally blooms in the fourth year from seed. A mature plant can produce a thousand blooms during the six-month flowering season from July to December and is usually at its best in September. It generally dies after about 10 years.

L. reflexum grows wild in the Cedarberg mountains of the south-western Cape and stands up to the cold fairly well.

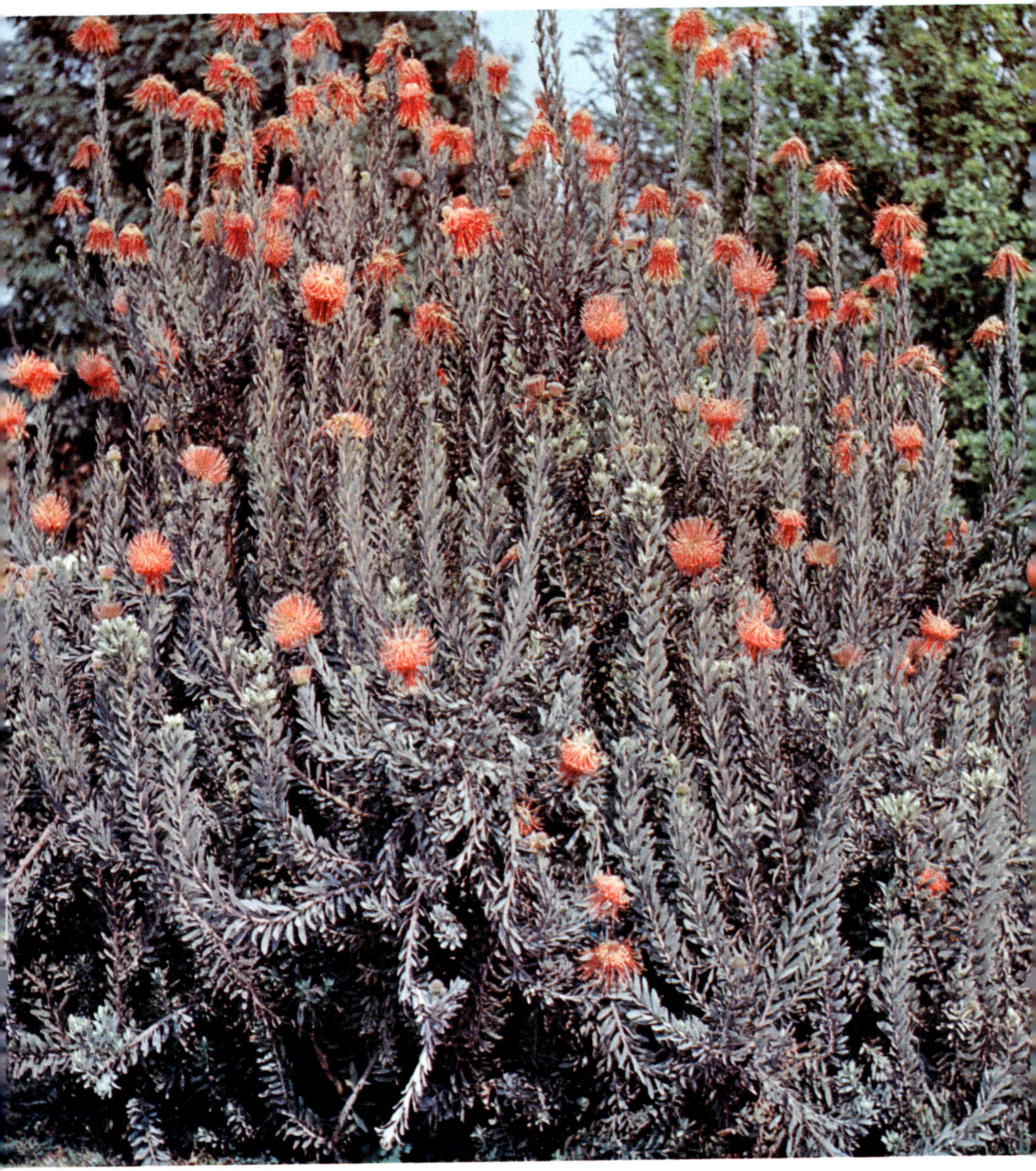

KREUPELHOUT, CRIPPLE-WOOD
Leucospermum conocarpodendron (L. conocarpum)

The bright golden yellow flower of this species is unusual in the Protea family, making it an acquisition for the vase and garden. It is not one of the showiest of shrubs, however, as the flower-head half nestles in the surrounding rosette of deep green leaves which clothe the stems luxuriantly. The cone-shaped flower-head, which measures almost 3 inches across, emerges more as it ages, for the leaves open out wider to reveal the whole pincushion. The leaves could, of course, be stripped for florist work. The flowers are borne at the tips of the younger branches from spring to midsummer.

Each leaf measures about $3\frac{1}{2}$ inches long and over an inch wide, being notched at the tip, like all *Leucospermums*. They seem to overlap one another, all along the thick woody stems, which appear to be thick and fairly clumsy. They sway and ripple in the breezes that blow on the coast, withstanding the often fierce gales near the shore.

Perhaps it was the wind that twisted and curved the branches of this tree as it grew, reminding the early settlers of a gnarled and distorted cripple, and giving rise to the common name of Kreupelhout, or Cripple-wood. It grows to a height of about 10 feet, with the shape of a tree rather than that of a shrub. Branching from the base, the lower branches are covered with rough bark.

L. conocarpodendron grows in profusion on the mountainsides of the south-western Cape, forming an important component of the thick vegetation that covers the lower slopes with evergreen thickets. It may be seen on the marine drives around the Cape Peninsula as well as in the mountains of the Caledon and Stellenbosch areas. The plants are freely obtainable in the trade. This is a valuable shrub for sea-coast gardens as it withstands the wind and high humidity, being drought-resistant at the coast.

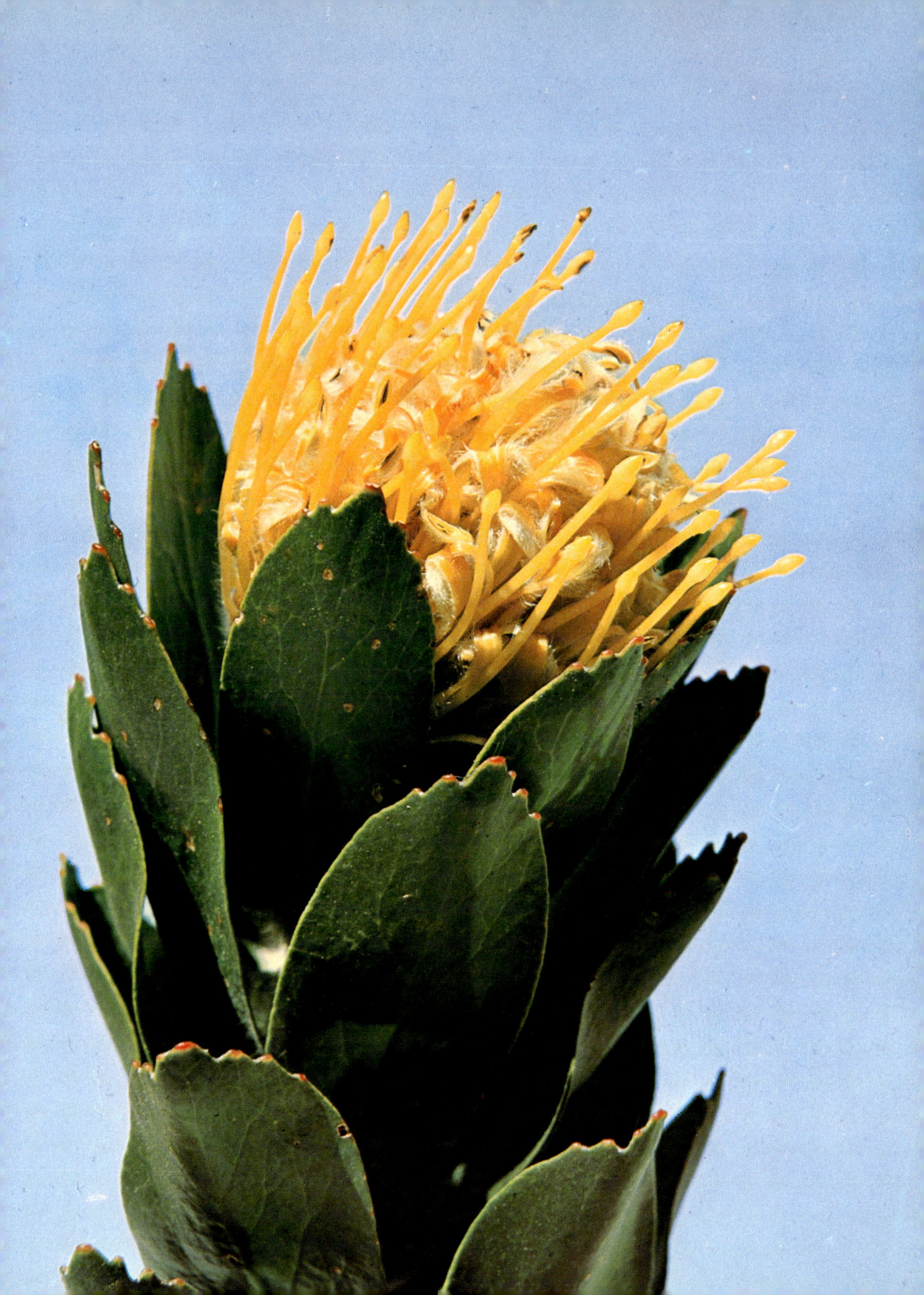

FIRE-WHEEL PINCUSHION, SPREADING PINCUSHION
Leucospermum tottum

Driving along the majestic Du Toit's Kloof Pass during summer, one is attracted by the neat rounded bushes of *L. tottum*, covered entirely with scores of dainty buff-salmon flowers. This species is valuable because it flowers later than most other Pincushions, starting in October and continuing until January.

The flower-heads are not quite as large as those of *L. cordifolium*, measuring about 3 to 4 inches across and having a flatter shape. The individual scarlet flowers are curled into ribbons forming a dome at the centre, while the styles or "pins" project stiffly for an inch beyond the flowers. The styles are straight in *L. tottum*, but are more gracefully curved upwards in *L. cordifolium*. Each style is yellow with a glossy pink or scarlet tip, creating the effect of a fire-wheel.

The bush generally grows to a height of 3 feet, but may grow to 4 or 5 feet, forming an attractive rounded shape. It becomes woody with age. The foliage is deep green, with each narrow tough leaf measuring about $2\frac{1}{2}$ inches in length and half an inch in width. They are not as deeply notched as most *Leucospermum* leaves and may even be entire at the tip. They are loosely arranged up the stems which reveal an attractive reddish tinge between the green leaves.

L. tottum grows in the S.W. Cape mountains in the areas around Worcester, Caledon and Ceres. It grows at fairly high altitudes and has been found at an altitude of 5,000 feet. This indicates that it is naturally cold-resistant. It likes regular watering during the winter months when it grows in the summer-rainfall area and should not be allowed to suffer from lack of water in hot climates.

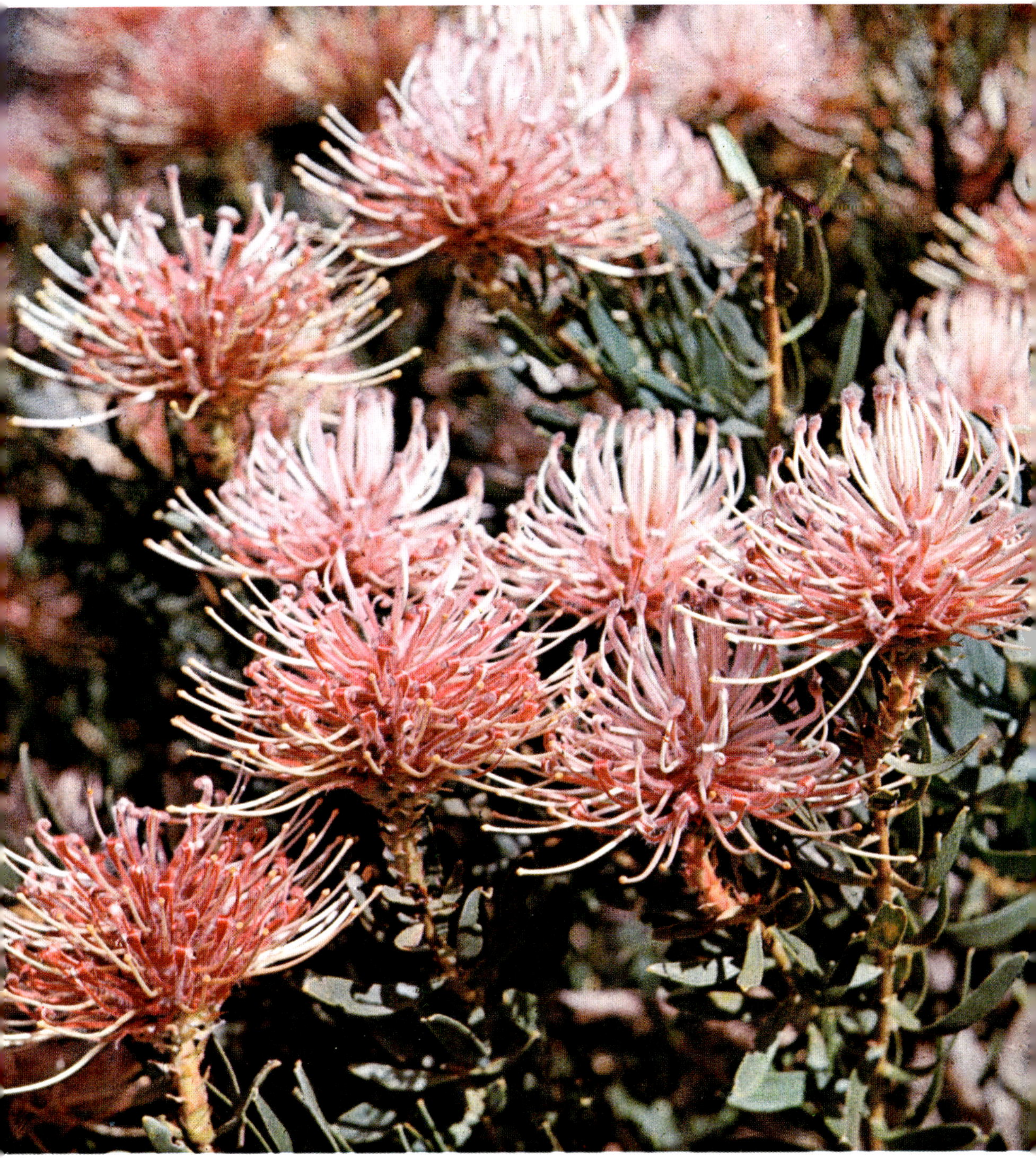

CATHERINE'S PINCUSHION
Leucospermum catherinae

The unusual champagne yellow colour of the flowers of *L. catherinae* make it immediately recognisable and attractive. The flattened flower-heads are very large, measuring at least 5 inches across, and when one looks down into the centre, the whole head resembles a spinning catherine wheel. Seen from a distance, the flowers have gracefully curved "pins", pink in the young stages, that seem to be "combed" to each side. As they mature the pins turn upwards and the flower becomes more tufted. The whole bush is lit up by these large incandescent pale lanterns of flowers that are arresting in the landscape. They are at their best in September and October and may still be in bloom during mid-summer. Some start flowering in winter.

The bush itself grows up to about 5 or 6 feet in height, requiring about 6 feet in which to spread. The flowers appear at the ends of the branches, but are not as densely packed as those of *L. cordifolium*. The large greyish-green leaves measure about $3\frac{1}{2}$ inches in length and an inch in width.

This is a robust shrub that comes from the mountains in the south-western Cape, and is quite cold-resistant. It grows equally well in the summer rainfall area of Kloof, Natal, where it experiences heat and humidity, as it does in the winter rainfall area of the Cape. It may die if it is hot and dry, but it is not a difficult plant to cultivate. It sets seed freely and is obtainable in the trade.

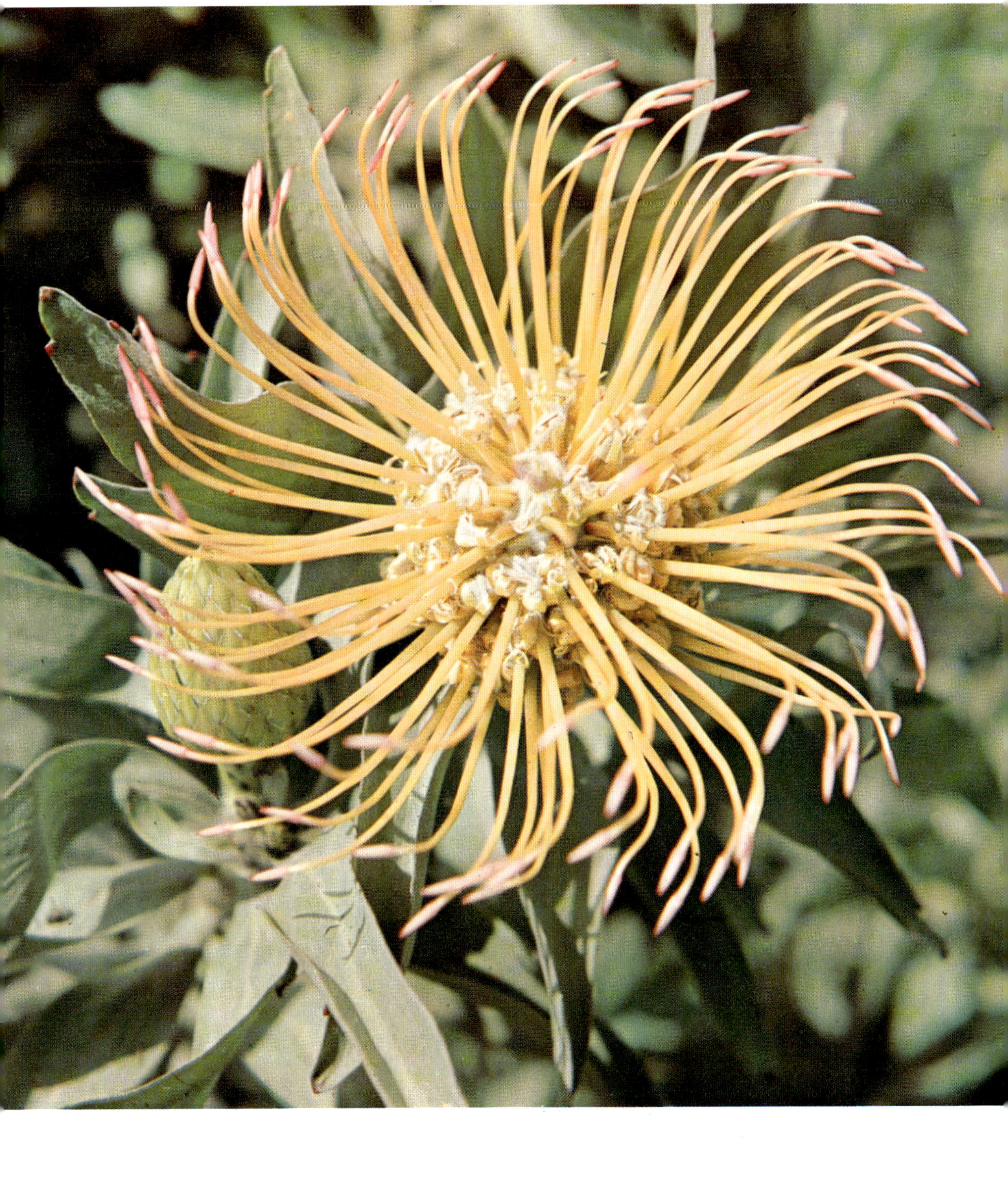

LARGE TUFTED PINCUSHION
Leucospermum praecox

A very showy landscape shrub, this is easily recognised by its bushy strong growth and the changing colours of the flowers. Other species like *L. grandiflorum, L. oleifolium* and *L. prostratum* also have this peculiarity, but *L. praecox* outdoes these in horticultural value as it is so free-flowering. It is obtainable in the trade.

The upright heads of flowers are reminiscent of those of *L. grandiflorum,* but are shorter, more compact and inevitably appear as twins at the tips of the branches in spring. They measure about 2 1/2 inches in length. They start as an orange-yellow, then the tips turn flame-coloured and finally the whole pincushion becomes deep crimson. The effect of all these colours on the bush at the same time is very eye-catching.

The foliage is luxuriant and the leaves overlap one another loosely along the length of the stems. Each leathery green leaf, about two-thirds of an inch wide and two inches long, is deeply notched across the blunt tip.

L. praecox grows easily in gardens and will grow in average well-drained soil. It needs very little attention and grows rapidly, flowering in the second year, when it reaches a height of about two feet. It spreads and grows up to about five feet, although small bushes often occur in nature. It is fairly drought-resistant, but should be watered regularly in winter, It withstands cold quite well and will tolerate alkaline soil, but does better in acid soil.

L. praecox is confined to the area between Albertinia and Mossel Bay in the Cape Province.

L. cuneiforme, previously known as *L. attenuatum,* is a similar, but more widespread species in nature, extending from Cape Agulhas eastwards along the coast to the Bathurst district. As it experiences such diverse climatic conditions in the wild, it should prove a tolerant species in gardens. It may be distinguished from *L. praecox* in that it has a persistent rootstock that allows it to coppice after veld fires and so produce multiple stems at the base, whereas *L. praecox* has a single stem. The base of the perianth in *L. praecox* is prominently swollen.

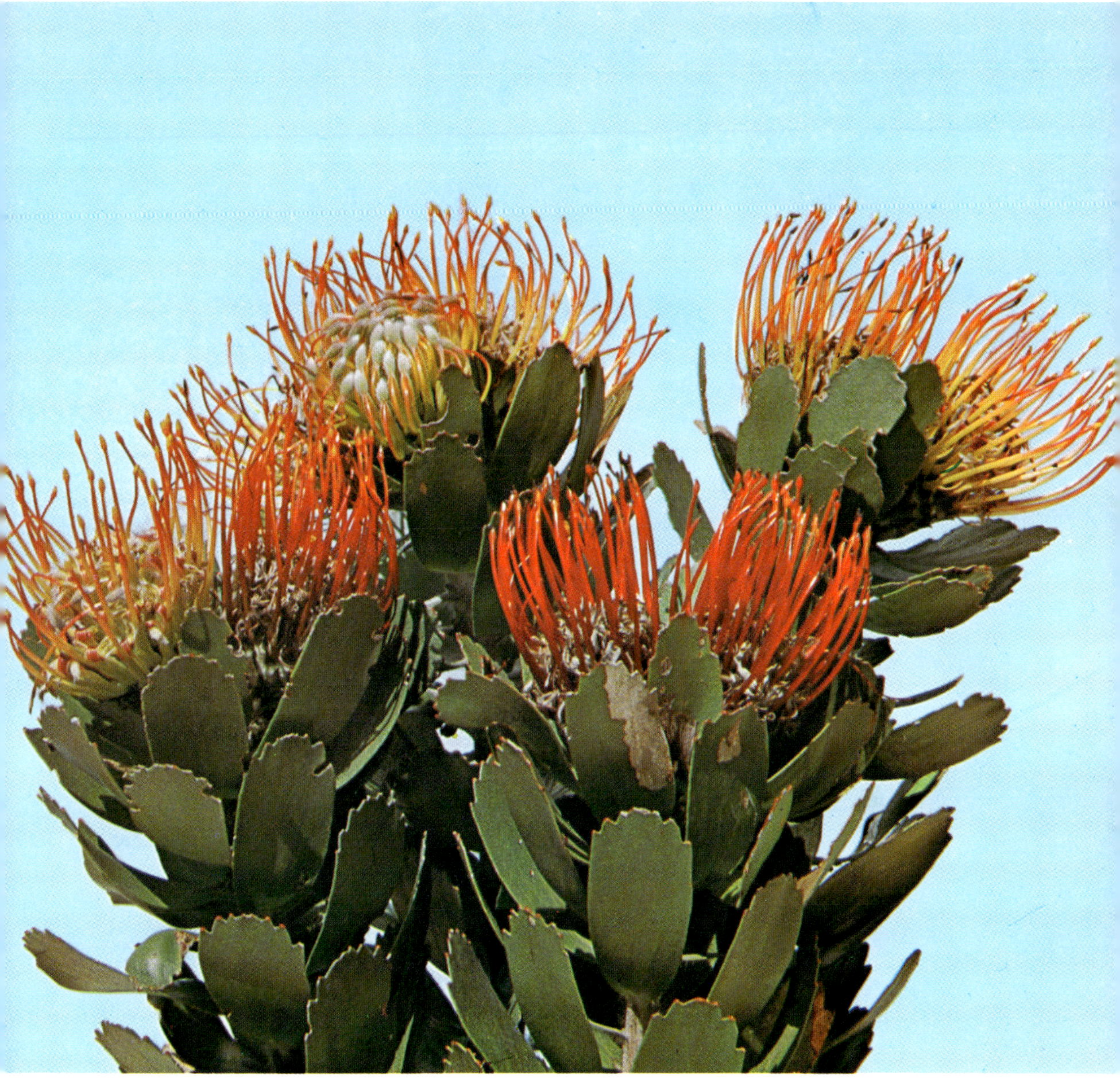

CREEPING PINCUSHION
Leucospermum prostratum

No visitor who walks up to the Proteas at Kirstenbosch can fail to be intrigued with the colourful little carpet of flowers formed by *L. prostratum*, draping over the rocks in springtime. The plant makes a mat spreading over an area of four or five feet and is heavily festooned with tiny flower-heads, measuring from $1\frac{1}{2}$ to 2 inches across. They are yellow when they first open, deepening to apricot and then red. As all colours appear at the same time, the effect is unusual and arresting. The flowers have a heathery fragrance of wild honey and bloom chiefly in the spring, while odd flowers appear almost all through the year. The small green leaves are slightly hairy and form a neat base for the flower-heads.

Every gardener who sees it wishes to grow it, but this is not easy. It needs exceptionally well-drained soil and grows in very sandy, acid soil in nature. A sloping bank, a rockery or a large window-box where drainage is very sharp, would be suitable. A warm situation with full sun is necessary, for it is tender to frost. Do not allow nearby plants to crowd it out of existence. *L. prostratum* grows slowly in gardens and may not flower for five or six years. Seed is rare and the plant is not obtainable easily. It grows wild in the south-western Cape on sandy flats near the coast in the Caledon area.

L. hypophyllocarpodendron (L. hypophyllum) is another creeping species, illustrated in *Wild Flowers of the Cape of Good Hope* by E. G. Rice and R. H. Compton. This grows flat on the sands in the Cape coastal areas and has yellow flowers, larger than those of *L. prostratum,* but not as decorative. It blooms from midwinter to midsummer.

L. spathulatum (L. ceresis), from the Ceres area, is a creeping pincushion with large red to pink flowers, similar to those of *L. tottum*. This is not obtainable, but might be worth growing as a spreading rockery plant.

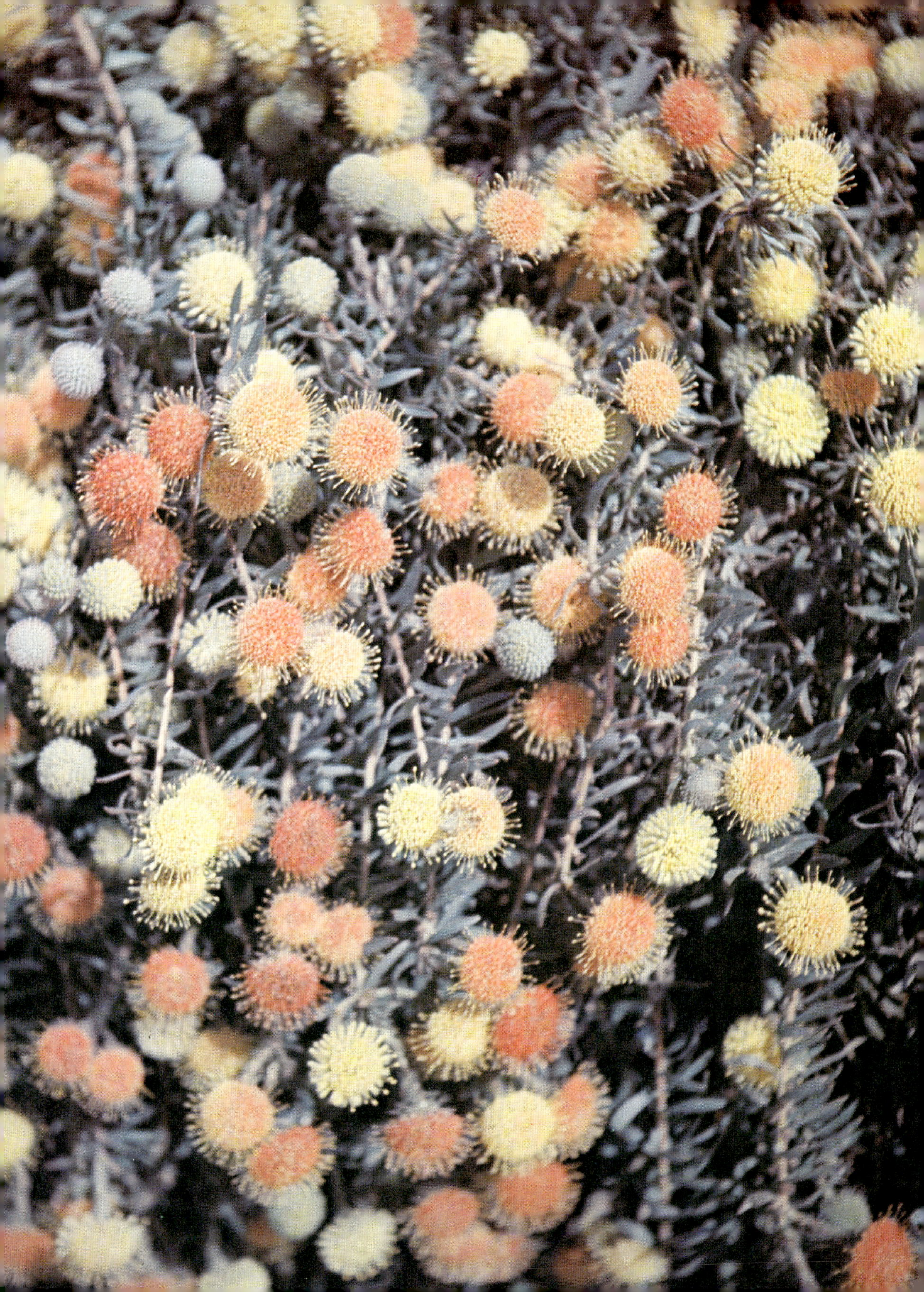

DWARF PINCUSHION
Leucospermum gerrardii

This small attractive shrub grows wild in the summer-rainfall area, flowering in September and October. It grows from 12 to 18 inches in height, springing from a thick underground root-stock and being replaced by new branches each season. It is frequently damaged by veld fires in nature, but would certainly prove an interesting plant for the rockery in the summer-rainfall area. It does not transplant well and should be grown from seed.

The dainty flower-heads measure about $2\frac{1}{2}$ inches across, and vary in colour. The showiest form is a deep glowing red, while there are also yellow forms. The narrow leaves grow up to three inches in length and vary from a quarter to one inch in width.

This Pincushion grows at high altitudes of 3,000 to 5,000 feet in the eastern Transvaal mountains around Barberton, in Natal and Swaziland. It generally grows in the open grassland amongst rocks, with very little other shrubby vegetation nearby. This is one of the three species of *Leucospermum* that grow exclusively in the summer-rainfall area.

The other two are *L. saxosum*, from the eastern Transvaal and the Chimanimani Mountains in Zimbabwe, and *L. innovans*, a new species described by J. P. Rourke, from the Transkei coast and southern Natal. *L. cuneiforme* also grows wild in the summer-rainfall area, but occurs plentifully in the winter-rainfall area.

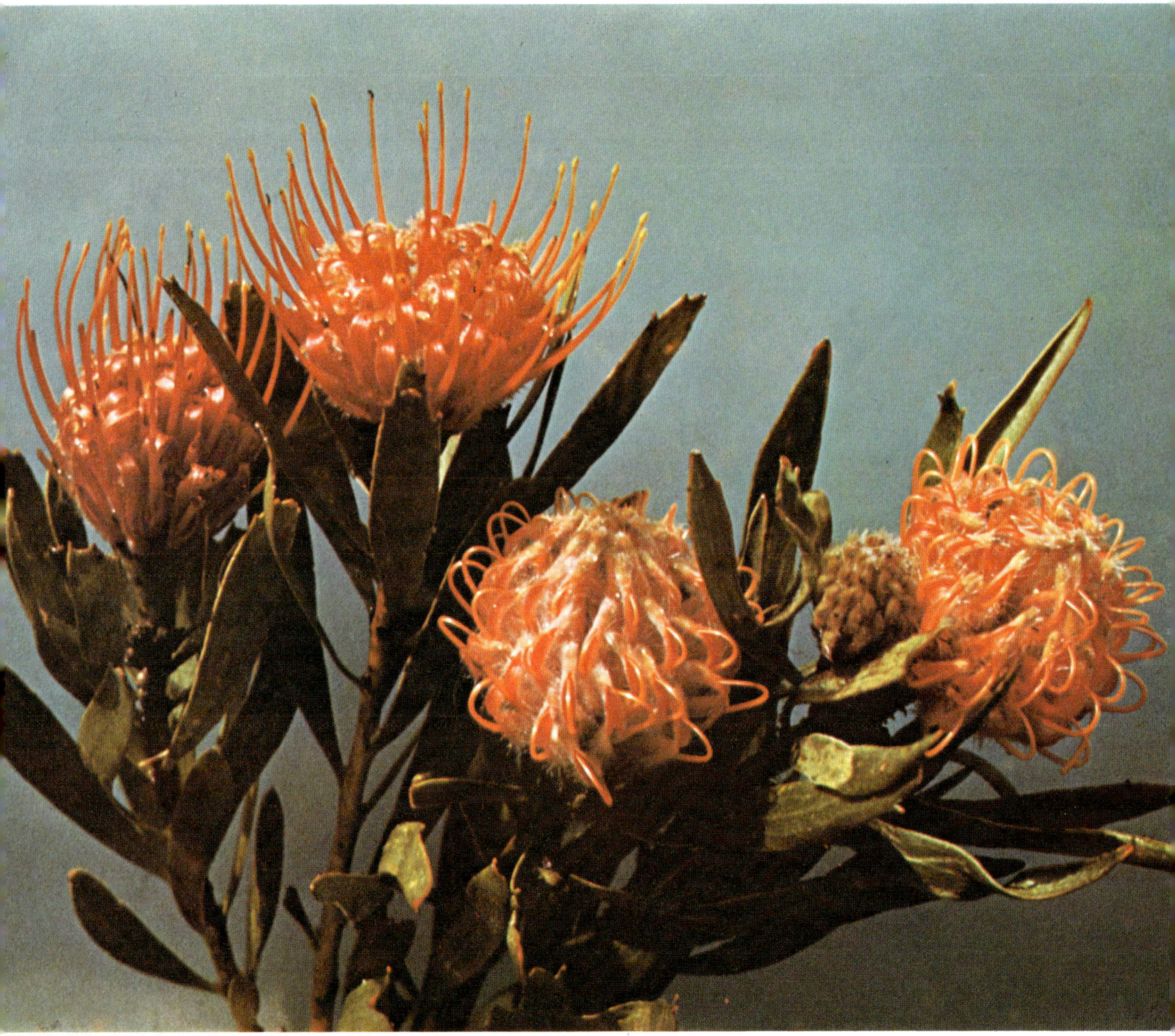

NARROW-LEAF PINCUSHION
Leucospermum lineare

This plant has the typical pincushion flower-head, measuring about 4 inches across, that one associates with the Nodding Pincushion, *L. cordifolium,* but it is easily distinguished by its attractive pine-needle-like leaves. These grow up to about $3\frac{1}{2}$ inches in length, pointing upwards all along the stems, and give the whole bush a light and delicate appearance.

The flowers of *L. lineare* have a rich flame-tangerine colour and gleaming texture. Some plants have paler flowers, but the deeper colour forms are, naturally, more desirable. They come true from seed. The flower-head is slightly more flattened than that of *L. cordifolium.* The flower-heads tip each of the numerous branching stems of the bush and are most decorative in the landscape. They bloom prolifically during September, and remain in flower until summer. The bush grows to a height of 2 to 3 feet and is suitable for any garden.

L. lineare grows wild in the Paarl and Fransch Hoek mountains of the south-western Cape. It is not easily obtainable in the trade, but does grow easily in the garden. It puts up with quite dry soil and endures very cold weather if its position is not too exposed. It starts flowering in the second or third year and will bloom when it is only about 18 inches in height. Watering should be given regularly during winter if it is grown in the summer-rainfall area. A west-facing slope in full sunshine is ideal.

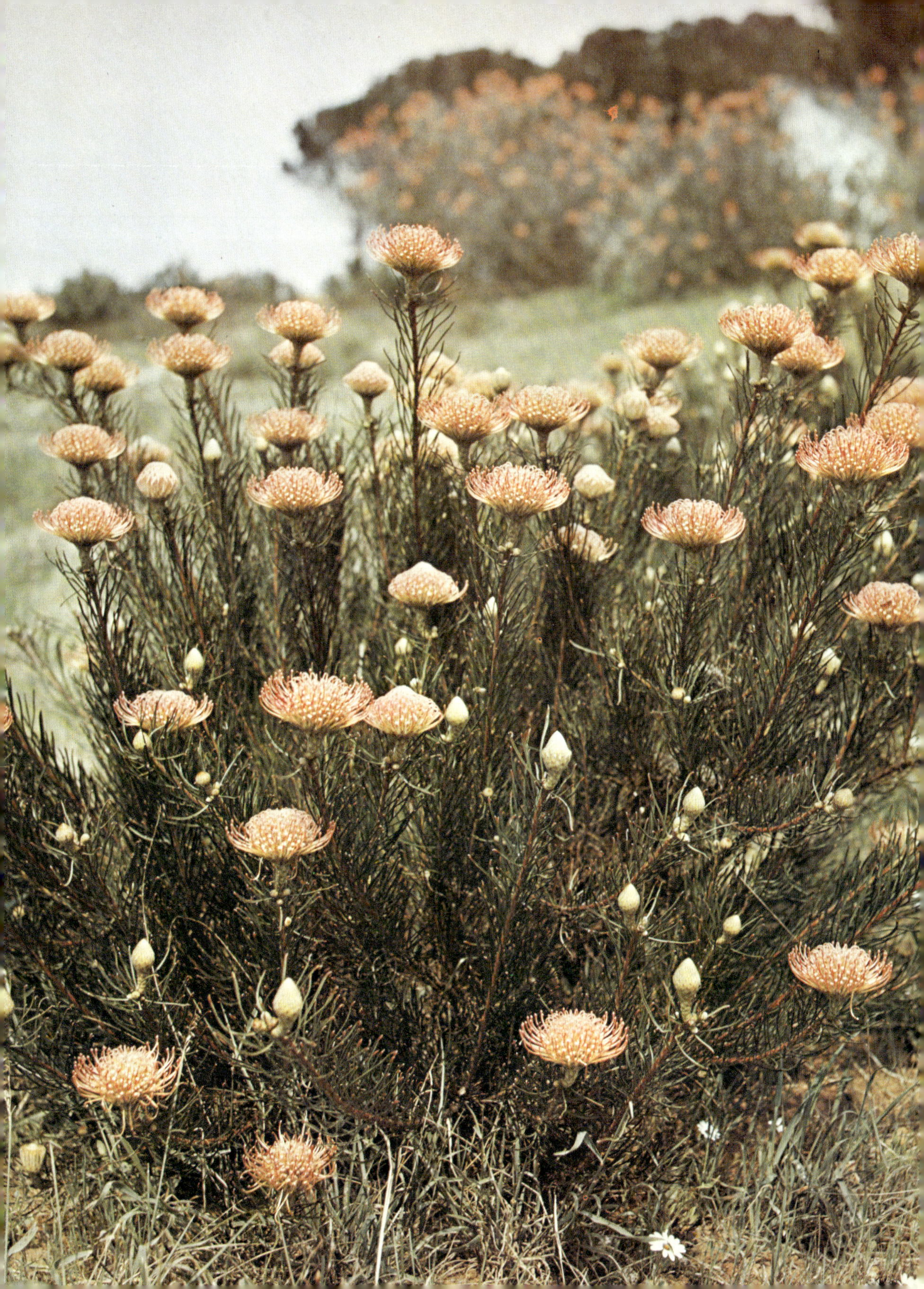

NOTCHED PINCUSHION, UPRIGHT PINCUSHION

Leucospermum vestitum (L. incisum)

One might say that most Pincushions have leaves that are incised or notched on the margin, but those of *L. vestitum* are very sharply cut into the tips. This species has a bright glossy flower similar to that of *L. cordifolium,* but it has a more flat-topped construction and the layman can readily distinguish it by this feature.

The flower-head of *L. vestitum* measures five inches across and the "pins" extend to a height of about $2\frac{1}{2}$ inches. The colour is a bright orange-red and the flowers are showy in the garden and in the vase. It should be grown in a collection of the more attractive types of Pincushion Flower. It starts flowering in midwinter and continues in bloom throughout spring, summer and autumn. The smooth green leaves are over two inches in length and almost an inch wide, with 2-4 deep notches at the tip. They overlap up the stems and make the shrub a luxuriant evergreen. The fairly upright bush grows to a height of about four or five feet.

This species grows easily in the garden, needing the same attention as *L. cordifolium*. It grows wild in the south-western Cape, in the Caledon area, near the Breede River as well as in the Worcester region on Mitchell's Pass and in dry areas near Tulbagh.

The species formerly called *L. ellipticum* and *L. medium* are now also placed under *L. vestitum*.

Leucospermum glabrum. Distinguished by its large, smooth leaves, with 7 to 14 teeth or notches at the top, this is a robust shrub that may grow to over 2 metres in height, with a single trunk at the base, The large oval "pincushion" measures 7-9 cm across and the flowers are bright orange-red, covered by densely woolly bracteoles when young. It blooms in spring and early summer and grows wild on the lower slopes of the Outeniqua Mountains near George and Knysna in moist peaty soil, in association with other tall shrubs.

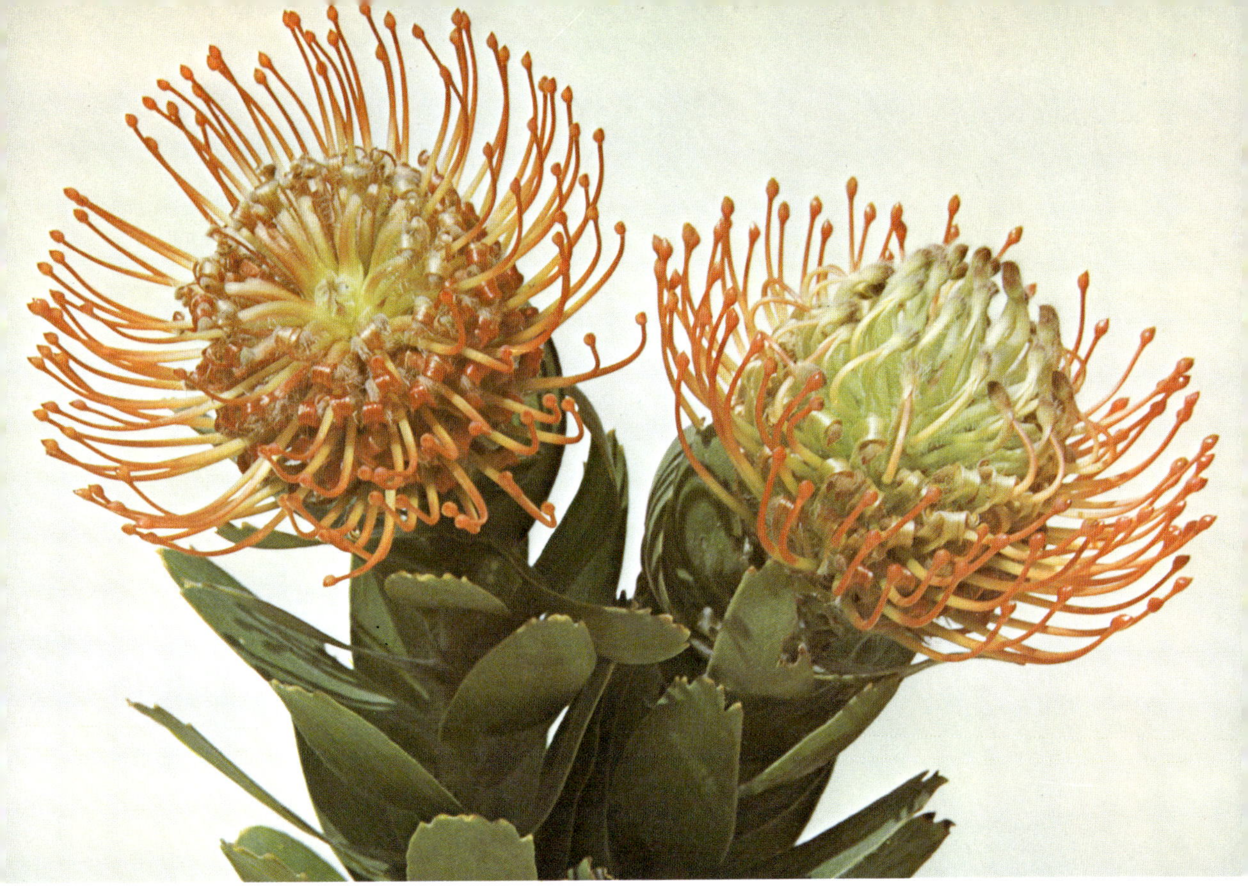

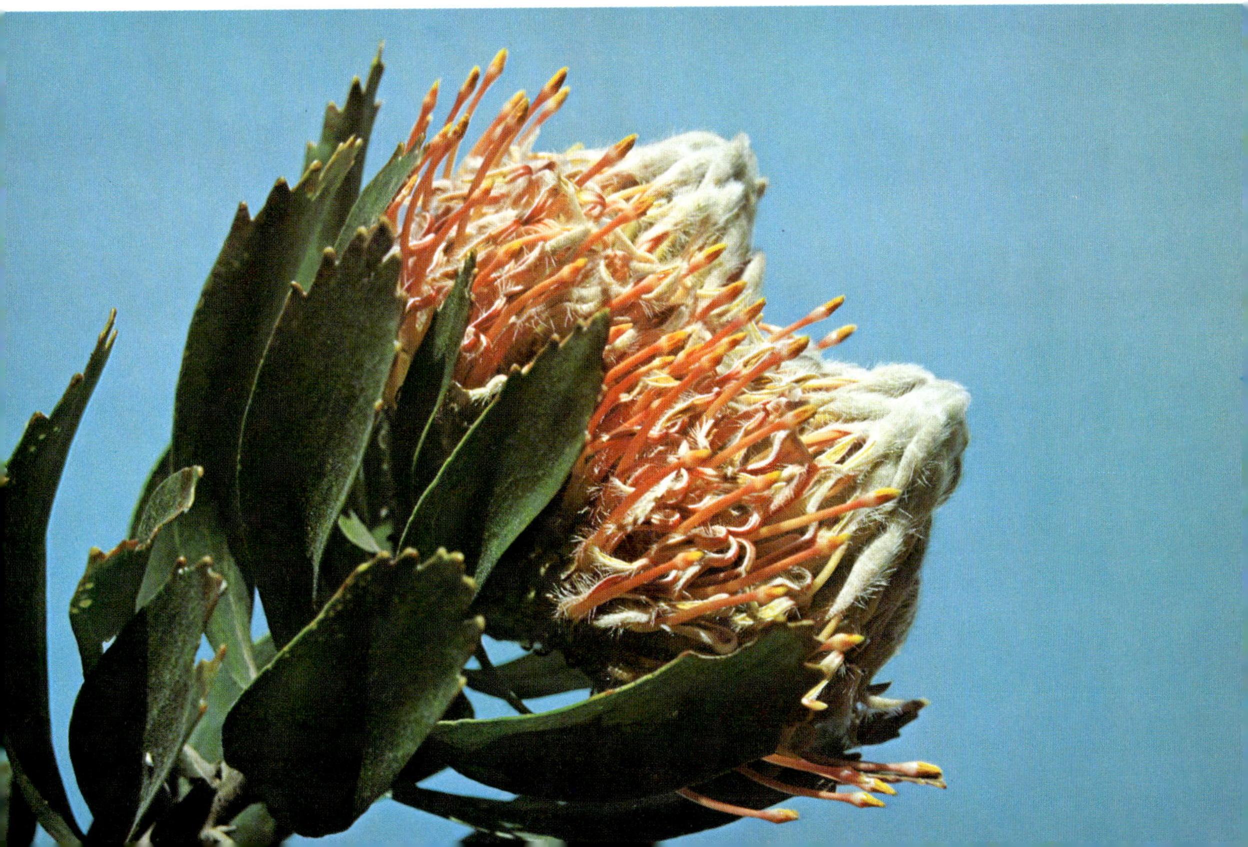

SMALL PINCUSHION
Leucospermum truncatum

A dainty plant for the small garden, this has prolific small Pincushion flowers that are pretty on the bush as well as in the vase. It is worth cultivating and is available in the trade.

The flowers appear in spring and early summer. Each flower-head, which measures about two inches across, is orange-yellow in effect. When one examines the flower one can see that the upstanding "pins" are orange or yellow, while the neatly curled "ribbons" of the flowers at the centre are yellow with red stripes. The flowers stand well above the foliage and show up clearly. They are thickly-hairy. The small greyish-green leaves are about 4 cm long and 8-15 mm wide. The fact that they are spaced at intervals along the stems gives this shrub its light and dainty appearance.

L. truncatum reaches a height of four feet and commences flowering when it is about 18 inches in height. It is particularly attractive in a rockery. It grows with ease in gardens, needing well-drained soil, a sunny position, water during winter and some protection from frost. It grows wild exclusively on limestone formations along the Bredasdorp coast in the Cape and will grow in soils up to pH 8.

L. muirii is similar to *L. truncatum,* but the flowers are only slightly hairy and the leaves are narrower, from 4-10 mm wide. This species grows on loose, acid drift soil in nature and is found only around Albertinia.

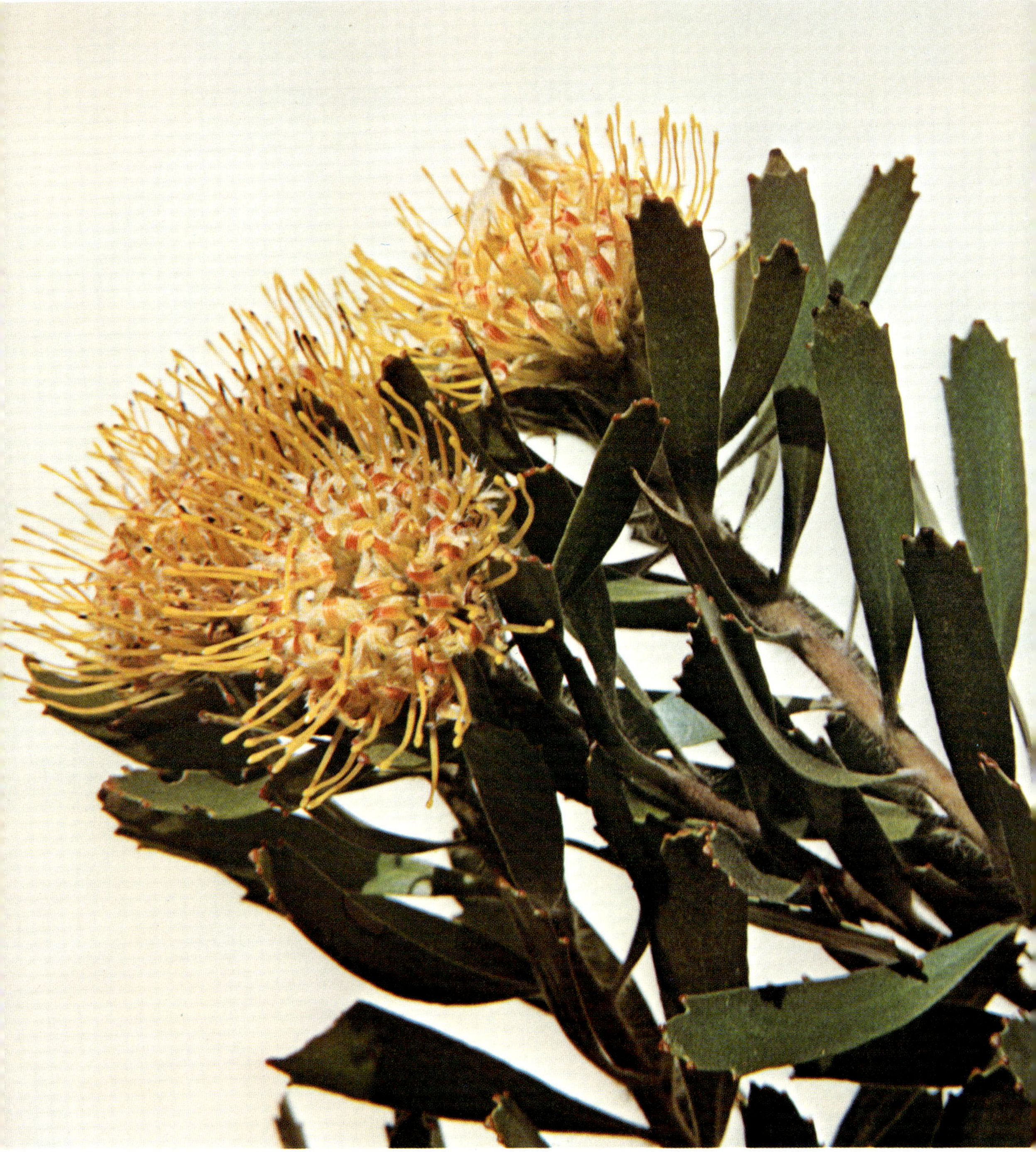

RAINBOW PINCUSHION
Leucospermum grandiflorum

The most arresting feature of this species is the colouring of the flower-heads, which give the effect of red and yellow flowers on the same stem. The flower-heads are tufted, measuring about 3 inches long and 2 inches wide, and two or three appear together at the tips of the branches. Each individual waxy flower is yellow, being tipped with glossy red as it becomes older. The whole flower-head slowly becomes orange and then crimson when mature. As the separate flower-heads are at different stages of maturity, the effect is given of red and yellow flowers blooming together. Each head elongates and becomes narrower as it ages. They bloom mainly in the spring.

The shrub grows to a height of about 6 feet in a fairly upright shape. The grey-green leaves are wide and taper to a blunted point, but may be tridentate at the apex. *L. grandiflorum* may be cultivated easily, requiring well-drained soil and water during winter. It endures cold weather well. This species grows wild near the tops of high mountains in the south-western Cape, in places near Klaver and Clanwilliam. Although its specific name implies that it is one of the best *Leucospermums,* it is by no means the grandest, but it deserves a place in a collector's garden and should be grown for interest.

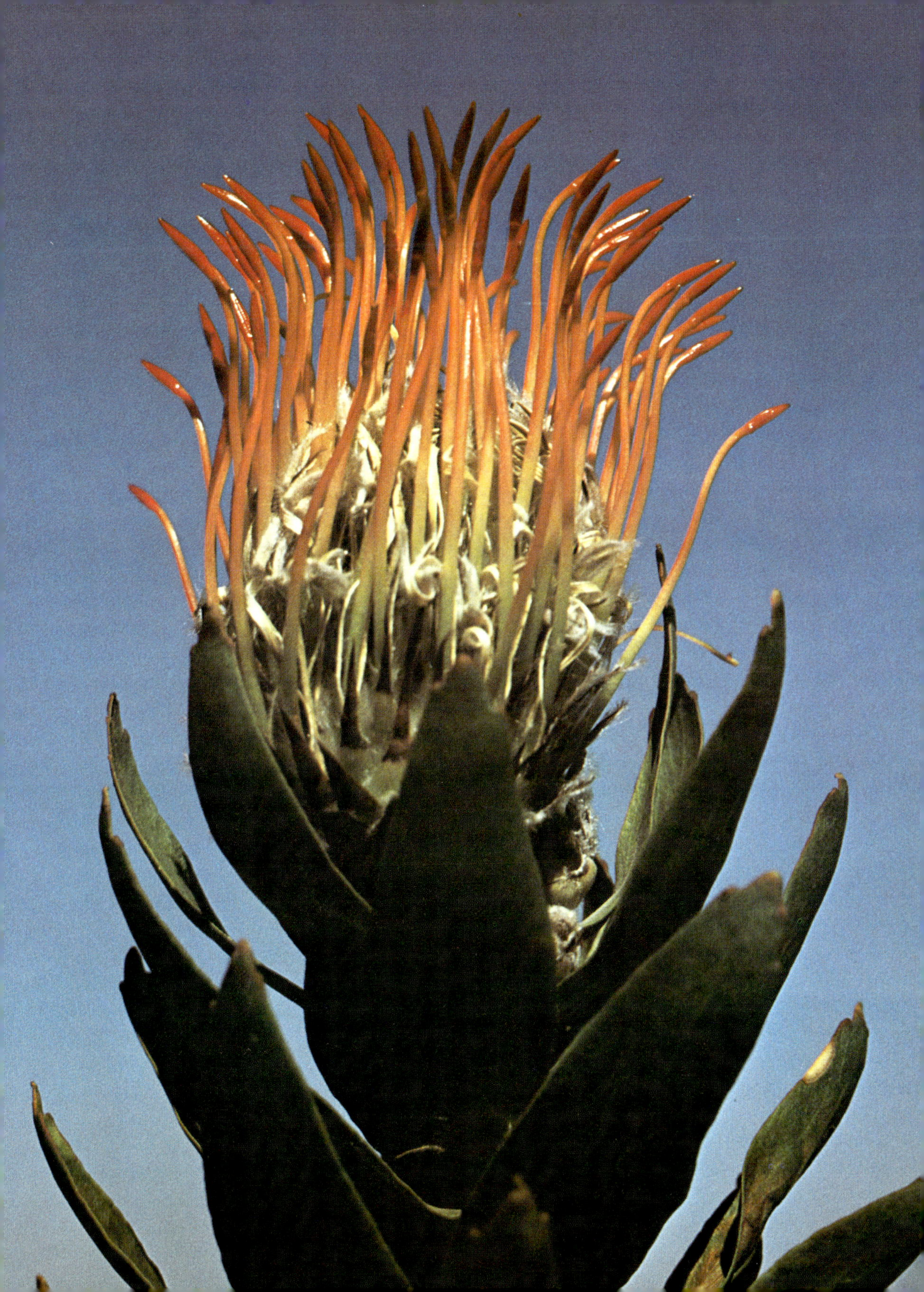

TUFTED PINCUSHION

Leucospermum oleifolium
(L. crinitum)

This species is like a miniature *L. praecox* in that it has tufted flowers that are yellow when they are young, changing to red as they mature. Twin or triple flower-heads grow together at the tips of the branches so that the different colours appear side by side, creating a variegated and colourful effect. The small flower-heads are at their best in the spring. They measure about an inch across and the "pins" are like tufts of soft hairs. The greyish, hairy leaves are generally about $2\frac{1}{2}$ inches long and $\frac{3}{4}$ inch across, as shown in the photograph. *L. oleifolium* generally grows to a height of three to four feet. It should be cultivated in the same way as other Pincushions from the winter-rainfall area. It grows wild in the south-western Cape near Caledon, Betty's Bay, Somerset West, Bain's Kloof and Worcester.

L. mundii is a similar species, that could easily be mistaken for a variety of *L. oleifolium,* which is found in the Tradouw Pass, between Swellendam and Barrydale. It bears six to nine tiny flowers packed together in rosettes at the tips of the branches and surrounded by very broad leaves, like a miniature posy. The leaves are oval and pointed, about $1\frac{1}{2}$ inches wide, and heavily notched along the top half. The flowers vary in colour through yellow, orange and red, being very attractive. (See opposite, below.)

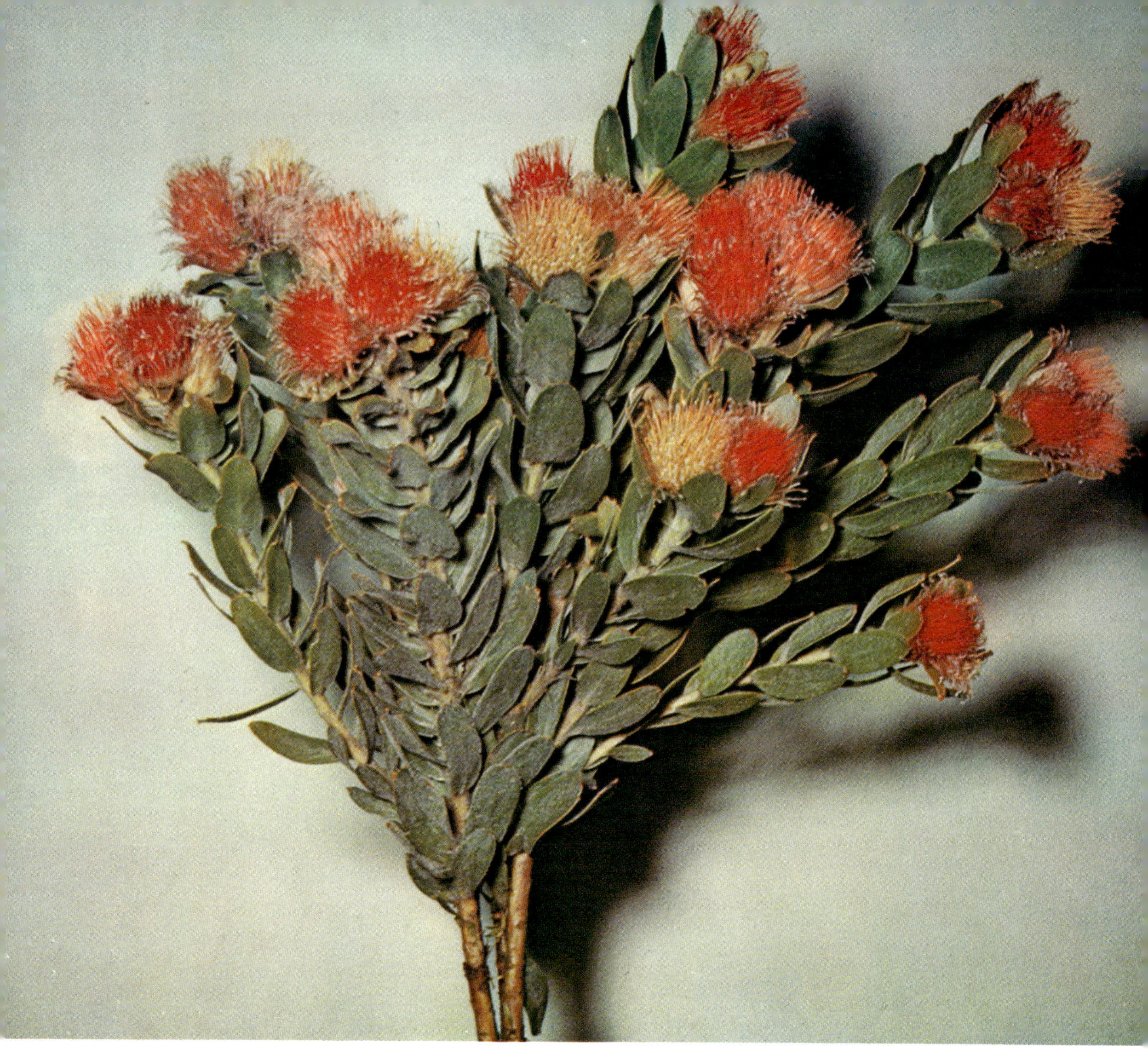

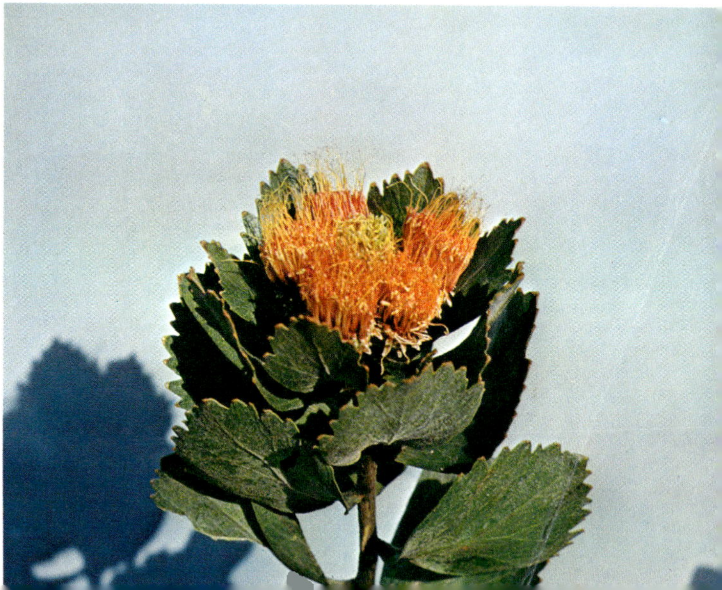

WHITE PINCUSHION
Leucospermum bolusii (L. album)

Pure white flowers are rare in Pincushions, so that this small-flowered species is of interest for this reason. It is used extensively in "mixed bouquets" of Cape flowers where it enhances more colourful blooms.

The flower-heads themselves are tiny, measuring less than an inch across, but are grouped together at the tips of the branches. Sometimes there are only two or three, but well-grown specimens may have as many as ten in one group. Each little head is like a flat-topped tuft of white velvet flowers, tipped with mauve. They bloom in spring and early summer, and have a scent of honey. The leaves are broad, measuring two inches in length and nearly an inch in diameter. This is a neat, bushy shrub that grows to a height of four to five feet.

L. bolusii grows wild in the winter-rainfall area of the south-western Cape. It grows easily in the garden, requiring the cultivation necessary for other Pincushions.

L. rodolentum is a species which is reminiscent of *L. bolusii*. It too, has small flowers grouped at the tips of the branches, but they are rounder and are bright yellow. The leaves are dove gray and this makes the plant an acquisition. It grows to four or five feet in height. This species also comes from the south-western Cape, growing wild from Darling to Clanwilliam and north to Worcester. Both of these species are available in the trade.

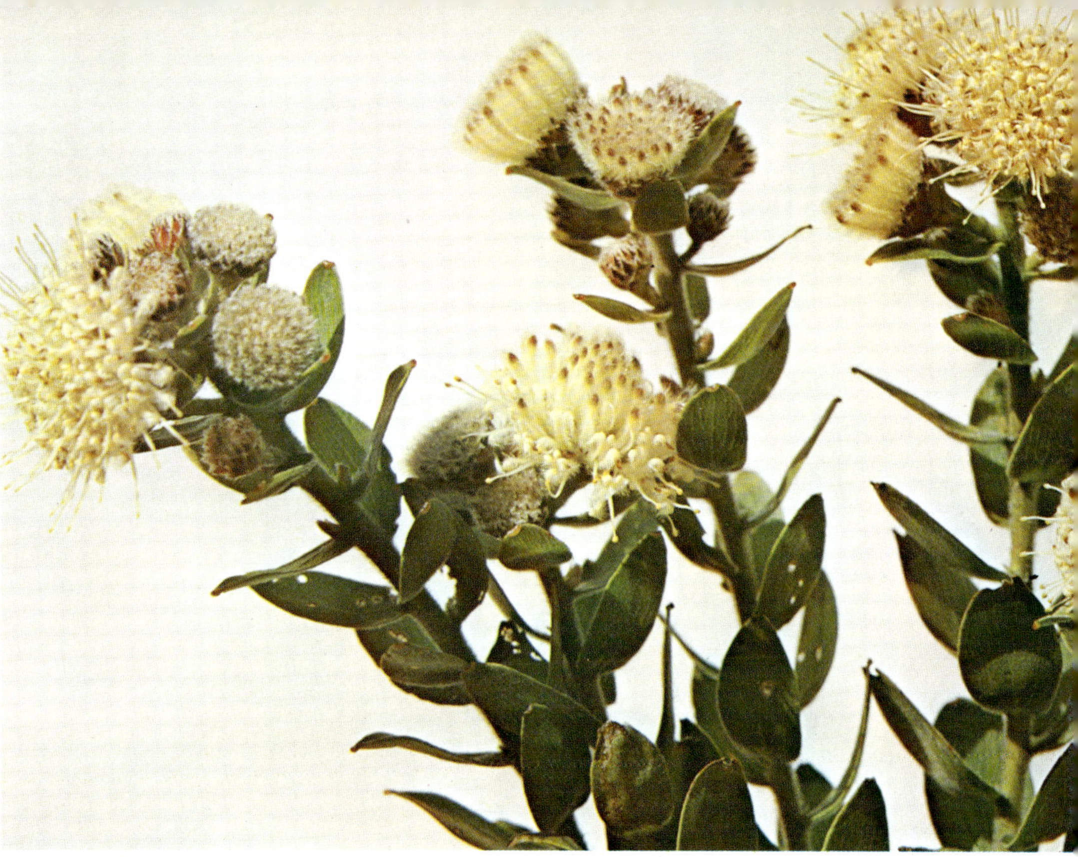

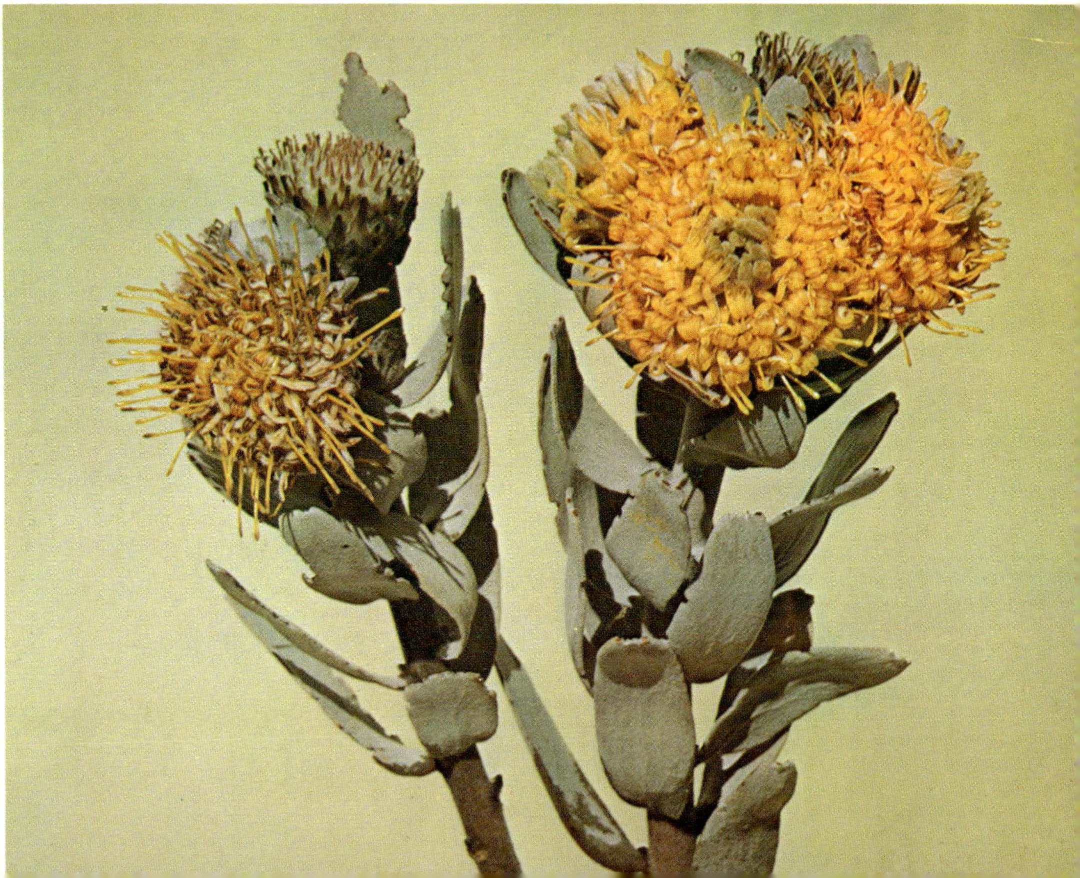

SILVER TREE, WITTEBOOM
Leucadendron argenteum

When breezes blow on Table Mountain at the Cape, there is no lovelier sight than the silvery branches of the Silver Tree, rippling and glinting in the sunshine. Young trees are most decorative, for their upright branches are thickly covered with overlapping greyish-green leaves, overlaid with fine, silvery hairs that catch the light. The branches are decorative in the vase, remaining fresh for over a week.

As the tree matures, the lower branches often become leafless, developing a rough bark. It ultimately reaches a height of 8 metres, but, as it often dies suddenly after 20 years, another can be planted to take its place and give renewed pleasure from the early stages. It is generally treated as a large shrub in cultivation, and should be pinched out while young to encourage bushy growth.

It is best to plant them when they are at least 25 cm in height. Losses frequently occur with smaller plants unless conditions are favourable. The Silver Tree grows easily, requiring well-drained soil and regular watering, especially during winter. It needs protection from frost in cold areas as the tips are burnt on large trees, while small ones may be killed altogether. A situation facing north and west is best in areas with a cold winter.

Silver Trees grow well in all parts of the country and may be seen flourishing in Natal, in the Kloof area, as well as in the mountains of the Transvaal and on the highveld. They grow wild in the mountains of the south-western Cape, especially on Table Mountain, and there is a magnificent grove on the lower mountain slopes at Kirstenbosch.

Male and female flowers, the former in the form of a fluffy yellow, pollen-bearing pompon of 4 cm across and the latter in the shape of a silvery cone, appear on separate plants in spring. The Silver Tree does not always flower, but generally blooms where the trees flourish in groves. A fungus disease frequently destroys Silver Trees at the Cape.

Leucadendron globosum

This rate, beautiful species from the Elgin-Grabouw area is well worth cultivation. It flowers in late spring and has keeled seeds. The large yellow male flowers measure about 5 cm across and have a strong lemon scent, being surrounded by broad yellow bracts. The smaller female inflorescence has a spicy smell. The oval, dark green leaves cover the sparse erect stems of 1-2 metres. (See opposite, below.)

As stated by Dr. Ion Williams, who revised the genus, there are 81 species of *Leucadendron*, without counting varieties and sub-species, including *L. olens*, a new species which he discovered in 1980. This was found in the George area of the Outeniqua Mountains and has small male flowers with a penetrating sweet smell. The erect bush of reddish hue has very tiny leaves.

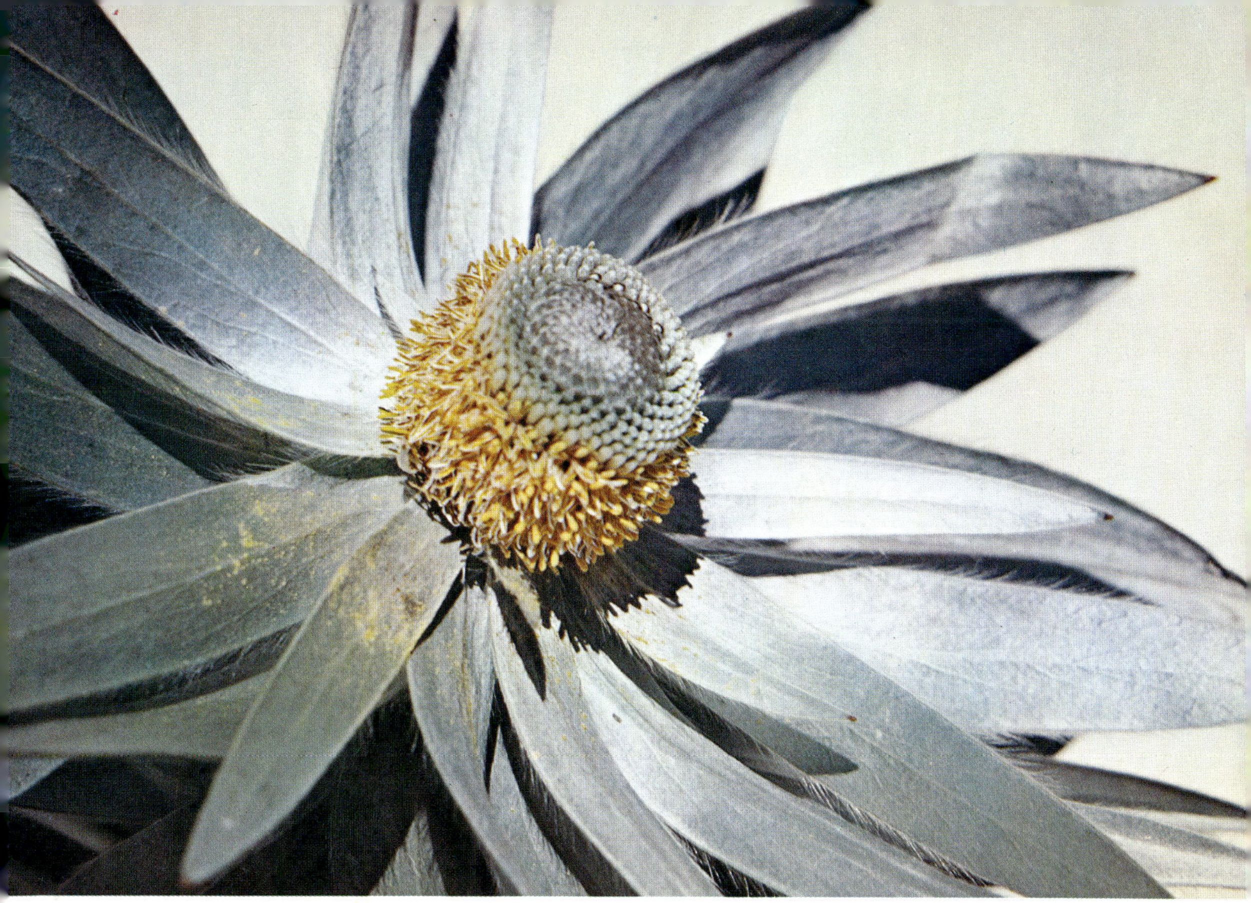

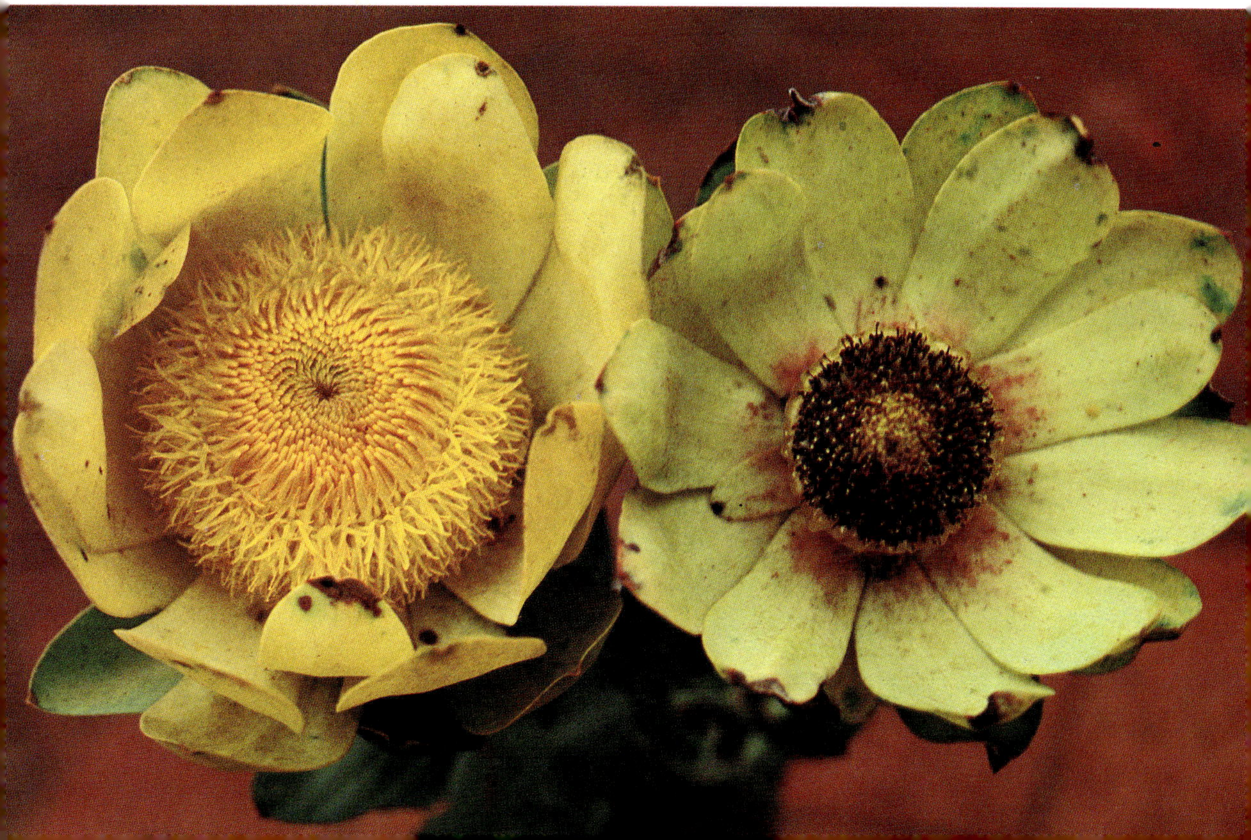

CERES GOLD-TIPS

Leucadendron arcuatum (L. crassifolium)

This outstanding species was exhibited at the mammoth Wild Flower Show held at Cape Town in September, 1963, to celebrate the jubilee of the National Botanic Gardens. There is no doubt that it would be a most desirable plant for the garden if it were made available in the trade. This possibility is very likely, for we have such knowledgeable and enterprising nurserymen who grow Proteas that what is rare today often becomes commonplace almost overnight.

The unusual feature of this species is the apricot colour of the topmost leaves or bracts, which resemble a tulip in form and size, colouring in the spring. This is a male plant, as indicated by the pompon of pollen-laden flowers at the centre. Leucadendrons have their male and female flowers on separate plants. *L. arcuatum* grows to a height of about 3 feet. It grows wild in the Ceres division of the south-western Cape. There are several other species with large bracts which are available in the trade.

L. tinctum, Rose Cockade, formerly *L. grandiflorum*, has broad upper leaves which are about 3/4 of an inch or more wide. They are pale green or creamy yellow, flushed with deep rose near the centre, from midwinter to spring. The male flower is a fluffy yellow pompon, while the decorative female cone is deep crimson and measures about an inch across. The female generally has the pale green leaves and the male the yellow leaves, but this is not always constant. This spreading shrub grows to about 4 feet in height. (Illus. frontispiece).

L. lanigerum is similar to *L. tinctum* in the way that it spreads, but grows to 5 or 6 feet. The leaves are tinged with autumn colours that vary on each stem, some being almost purple. The male flowers are yellow.

L. daphnoides has narrower leaves than *L. tinctum,* but they sometimes become altogether red. This species, which grows to 3 feet, is found near Paarl and Fransch Hoek.

L. microcephalum (L. stokoei) has yellow bracts with a striking brown ring around the florets and grows to 4 feet.

L. laureolum (L. decorum) has bright yellow bracts and the female has green cones. The bush grows to 6 feet and is common from the Peninsula to Caledon.

L. gandogeri (L. guthrieae) grows to 4 feet. The male has broad bracts, about $\frac{3}{4}$ of an inch wide and $2\frac{1}{2}$ inches long. They are yellow, shading to tawny, pinkish colours near the tip and there is a fluffy yellow centre. The female has larger yellow, pink-edged leaves up to 1 inch wide and $3\frac{1}{2}$ inches long, with an elongated cone. It grows near Caledon, Betty's Bay and the Hottentots Holland mountains.

L. sessile (L. venosum) has bright yellow bracts and centres, but there is also a dark red variety from the Gydo Pass. It grows to about 3 feet in height (See opposite, below)

In reviewing the few species above, one may realise how difficult it is to distinguish between the numerous species of *Leucadendron*.

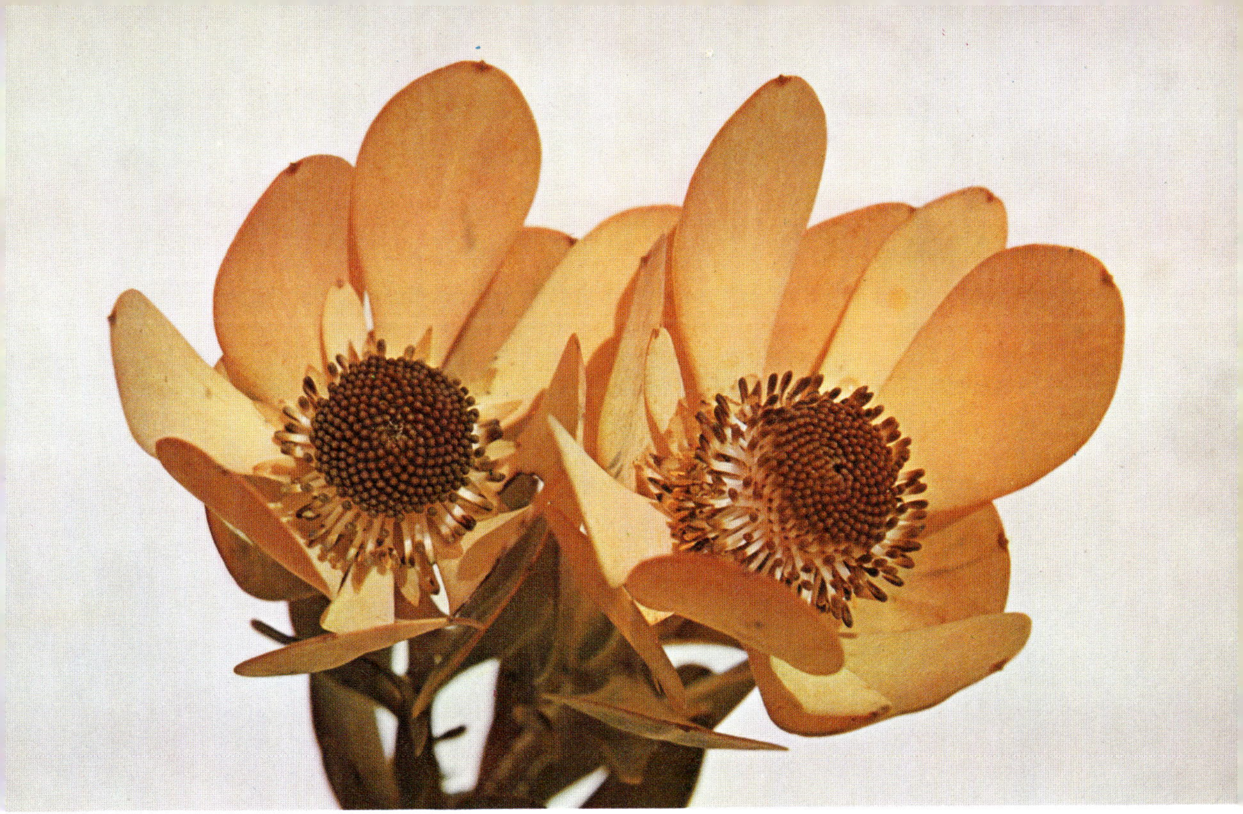

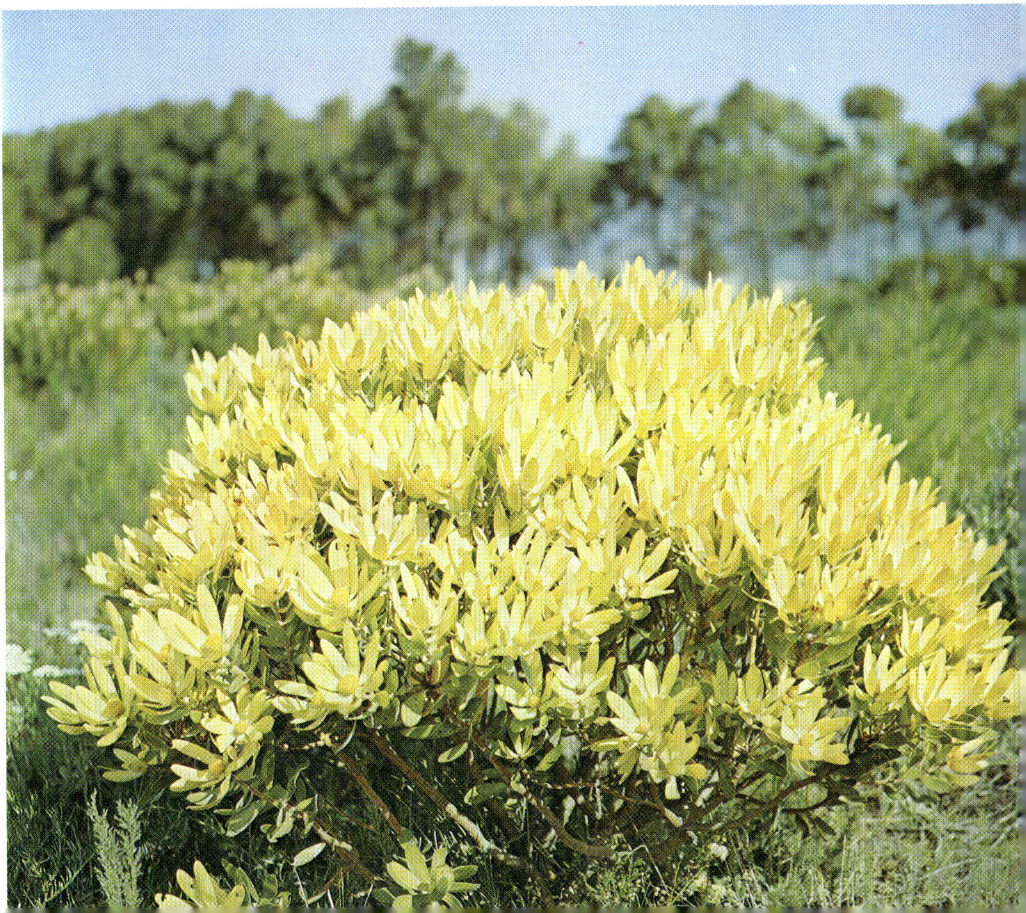

GOLD-TIPS, GEELBOS

Leucadendron xanthoconus (L. salignum) and *L. salignum (L. adscendens)*

In springtime at the Cape, around September, the hillsides are painted gold with *Leucadendron* bushes, particularly in the areas near Betty's Bay, Caledon, Stellenbosch, Piquetberg and eastwards to Uniondale. They are generally undistinguished shrubs in summer and autumn, but begin to colour in winter, becoming brilliant and showy in spring and early summer. The topmost, enlarged leaves are most colourful, but the whole bush, and more especially the top half, often changes colour. This colouring does not take place in every species, but the coloured ones should be grown for their late winter and spring foliage in the same way as some plants are grown for their autumn foliage. They are valuable for mixing with the more precious Proteas and Pincushions. Some retain their colouring even when dry and liven up a dried arrangement, retaining their starry quality for years in the vase. The two species illustrated here are wide-spread in the south-western Cape, lighting up the landscape in their thousands. They are very difficult to distinguish from one another, and a further complication in differentiating between them is that the name *L. salignum* is now given to that species previously called *L. adscendens*. The top photograph shows the male flowers of *L. xanthoconus* and the bottom picture, the female cones of *L. salignum*. In both species the male flowers form fluffy clusters of pollen-laden flowers and the female plants have small olive shaped cones and, sometimes, broader leaves.

L. xanthoconus has bright, lemon-yellow leaves near the top of the stems, which are larger than the lower leaves. The top-most leaves grow up to 2 inches long and less than a quarter-inch wide. They are very pointed. The male flowers are fluffy and yellow. *L. xanthoconus* grows into a bush of four to five feet in height, and is very similar to *L.salicifolium,* which is much larger, growing to 10 feet in height. *L. salicifolium* grows in very wet soil in nature, often at the side of streams, and is wide-spread in the south-western Cape. The female has an olive-shaped cone which is ridged and shaded with burgundy on the side facing the sun.

L. salignum has sharply pointed leaves that are generally shorter and narrower than those of *L. xanthoconus,* but the females generally have broader leaves, so that it is difficult to differentiate between the two species by means of the leaves. *L. salignum* is generally a much shorter bush than *L. xanthoconus,* growing from 1 to 4 feet in height, and makes a neat edge to a shrub border. The colouring of the foliage is brilliant and variegated. Some of the topmost leaves turn bright scarlet, others brilliant yellow or yellow tipped with apricot; some are crimson, burgundy or almost purple. They differ in the different districts in which they are found. Near Stanford in the Cape they resemble small, 15-inch tufts on the ground, while in the Worcester-Ceres area they grow to 4 feet and are a vivid scarlet. Stronger colours seem to be produced where soil and climatic conditions are favourable.

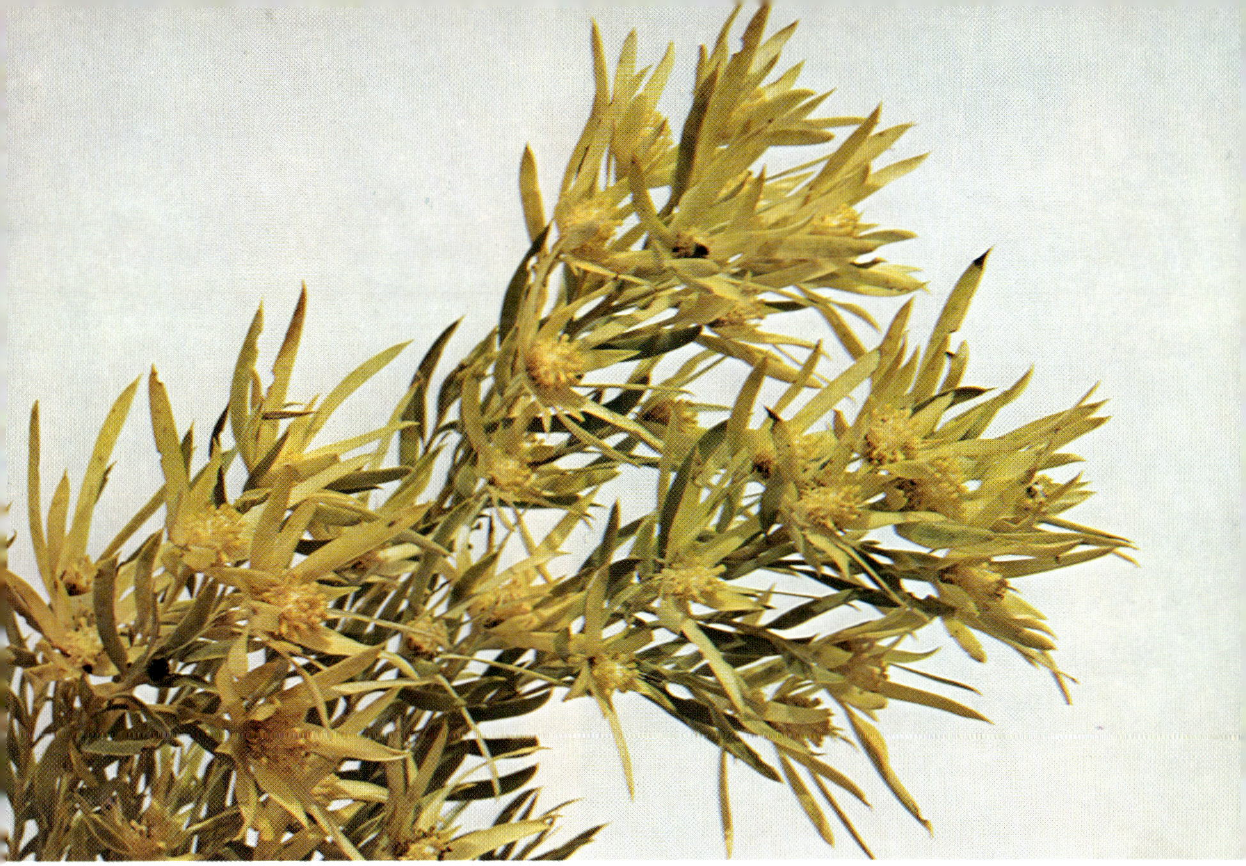
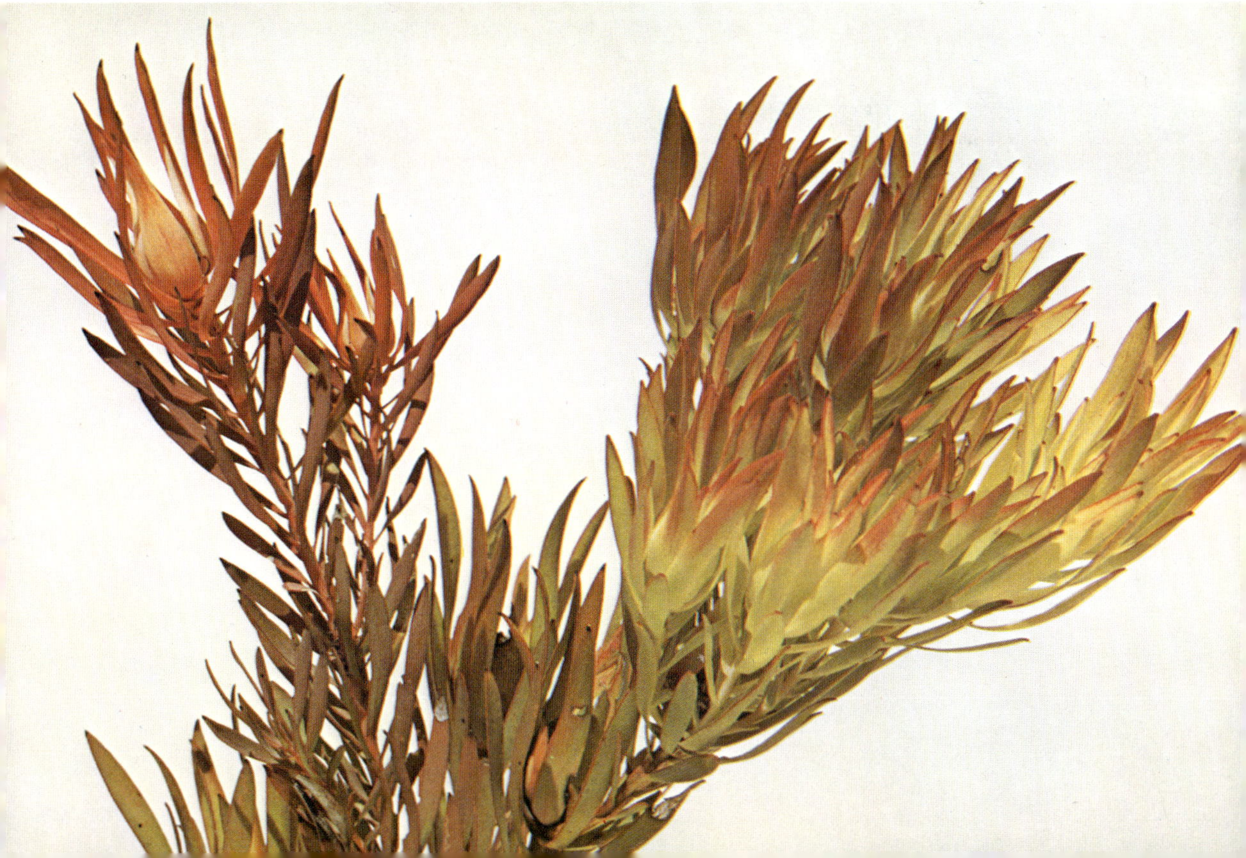

SUNSHINE BUSH, FLAME GOLD-TIPS
Leucadendron discolor

To see a hillside covered with male and female plants of *L. discolor* is, indeed, to see a hill of sunshine, because they are so cheerful and bright. This is one of the most decorative of all *Leucadendron* species, growing generally to a height of about seven feet. The erect branches end in broad, oval, tulip-like leaves that change colour in the spring. They are about two inches long and half an inch wide. The ordinary green leaves that grow up the long stem are a longish-oval shape, rimmed with red.

The male plant is more conspicuous than the female, for the upper leaves turn golden yellow, tinged with red at the edges. They open wide to reveal the showy male flower at the centre, which is shaped like an olive that is yellow at the base and bright red near the top. On the female plant the yellow cones are green, tinged with red and yellow.

L. discolor is available in the trade and grows easily in gardens, where it is becoming a favourite. It should be given well-drained soil with regular watering during winter, although it is a fairly drought-resistant plant. It needs a sunny position and should be given a warm situation in areas with cold winters. This species grows wild in the Piquetberg area of the south-western Cape.

L. eucalyptifolium is an extremely attractive bright yellow species that is available in the trade. It has long, graceful leaves that taper to a slender point, but vary in length. They seem to form an onion-shaped tuft at the tip of the branch, before flaring into an open star. The female has red cones. This shrub grows to a height of about 6 feet. It is wide-spread in the Cape, extending from Hout Bay to Tulbagh and Touws River in the north, and from Caledon eastwards to Port Elizabeth.

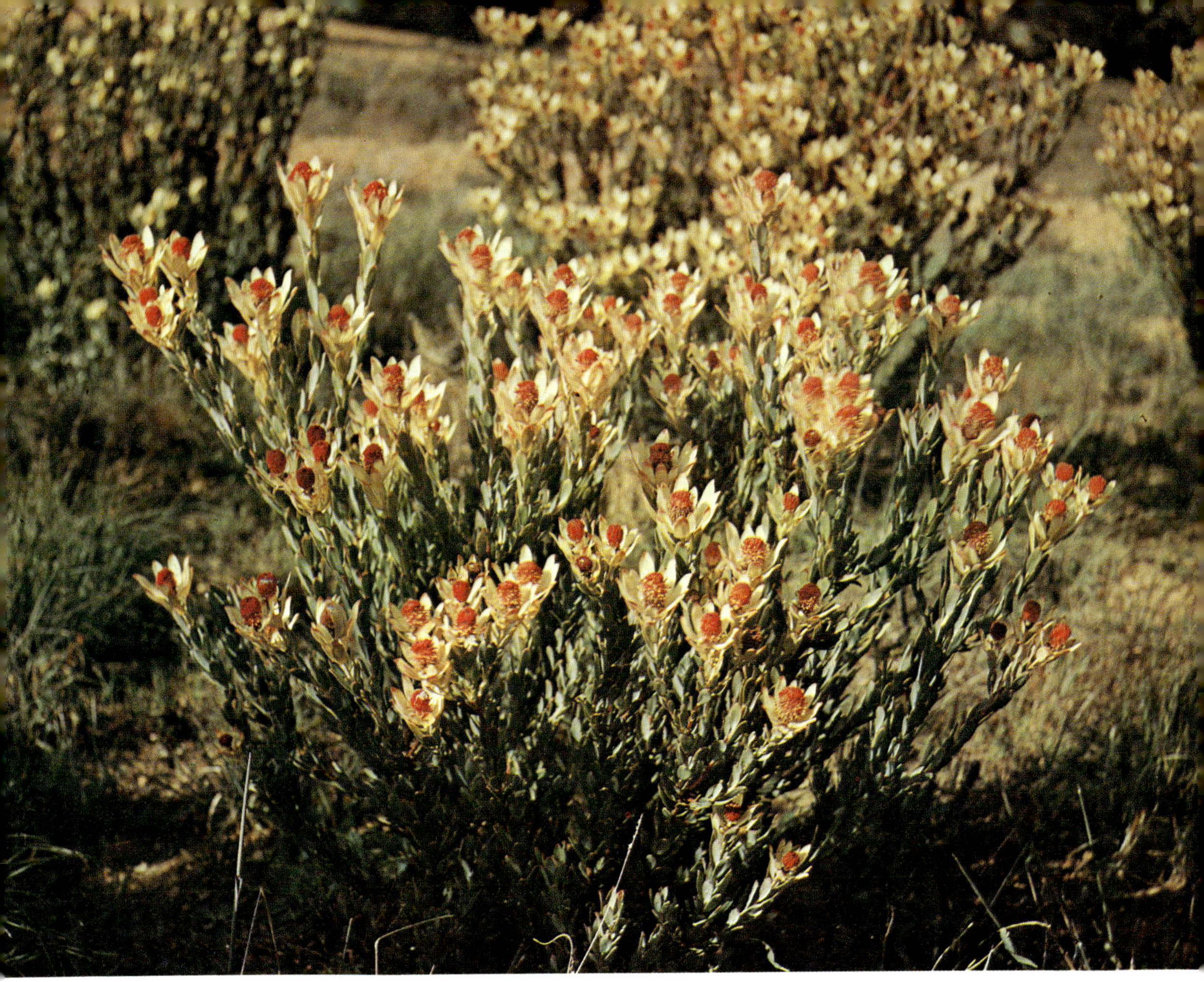

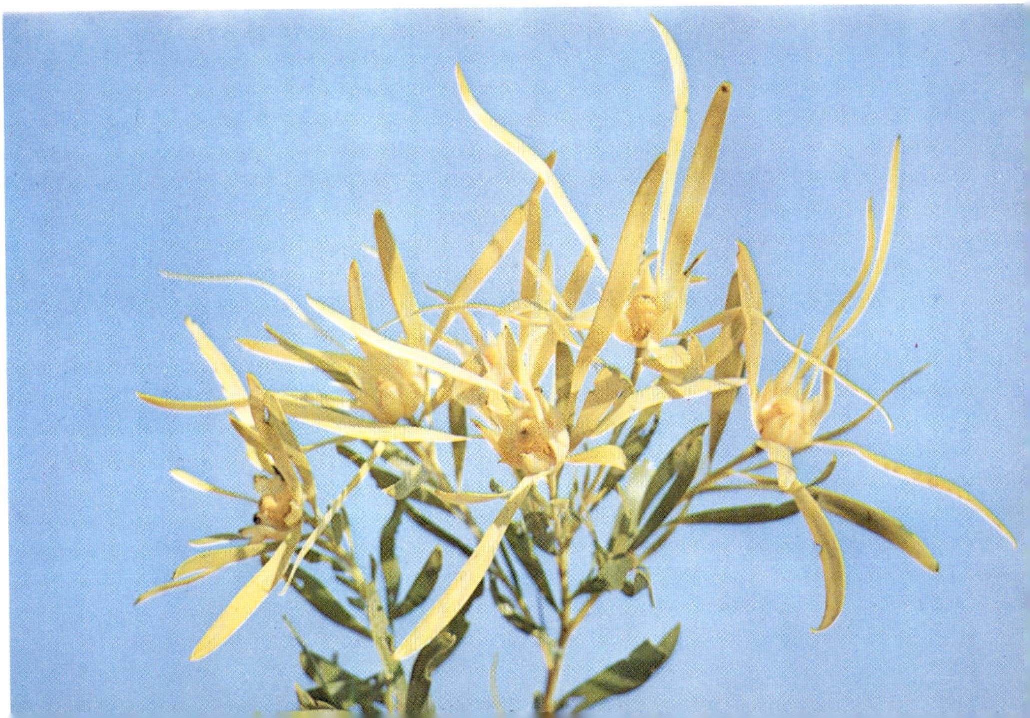

LARGE PINK-CONED LEUCADENDRON
Leucadendron coniferum (L. sabulosum)

The fact that this is such a commonly seen species makes it worth describing. It is prolific on the hillsides of the Cape Peninsula and occurs eastwards past Stanford to Cape Agulhas. As such, it is one of the plants frequently seen at the florists during mid-winter and early spring, when flowering material is especially appreciated.

The attractive feature of *L. coniferum* is the rosy, pink-tinged cone of the female plant. This is olive-shaped and measures about $1\frac{1}{2}$ inches in length, being ridged in waves. The leaves do not change colour, remaining grey-green and leathery. They measure about two inches in length and about a quarter-inch in width. The bush grows very large, varying from four to ten feet or more.

This species grows easily in gardens and is interesting because it will grow in alkaline soil as well as acid soil. It is available in the trade. (See photograph opposite).

SILVER-CONED LEUCADENDRON
Leucadendron muirii

This species grows exclusively in the limestone hills from Bredasdorp to Still Bay in the south-western Cape. It is, therefore, of special interest to gardeners who have alkaline soil in their gardens and find it difficult to grow the attractive members of the Protea family that need acid soil.

The silvery cone of *L. muirii* is most attractive and unusual. It is slim and olive-shaped, measuring about two inches in length, and gleams as if covered with metallic silver paint. As the cone dries and opens, it loses its silvery appearance, but traces of silver remain on the edges. The thick, leathery leaves are quite distinctive, being about $1\frac{1}{2}$ inches long and shaped like an oar, narrowing at the base, and about one third of an inch at the widest part. The leaves are similar in both sexes. The male flowerhead is also olive-shaped but much smaller and is not decorative. This evergreen shrub grows to a height of 3 to 4 feet.

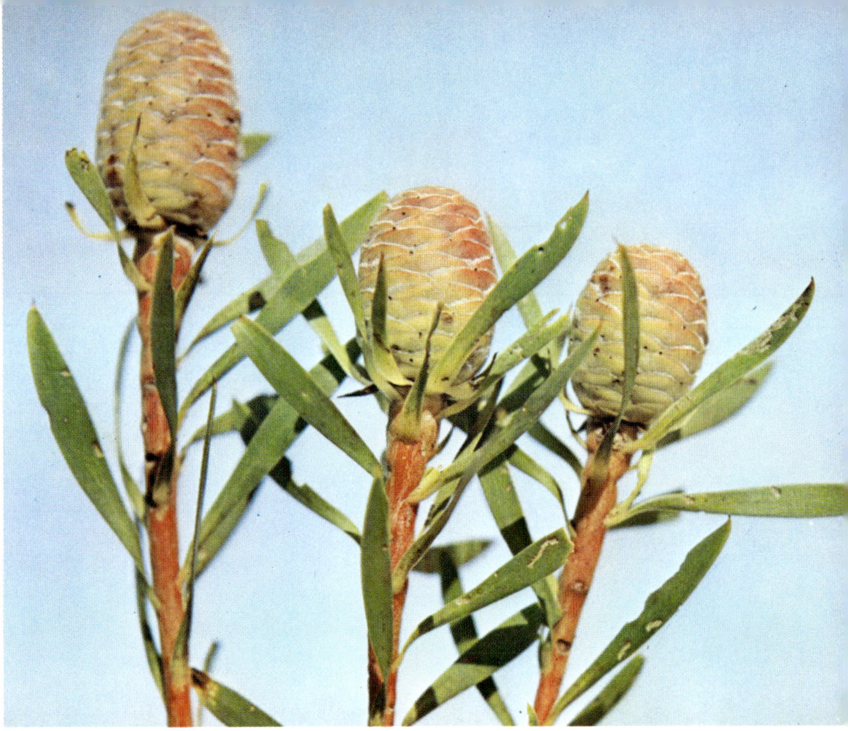

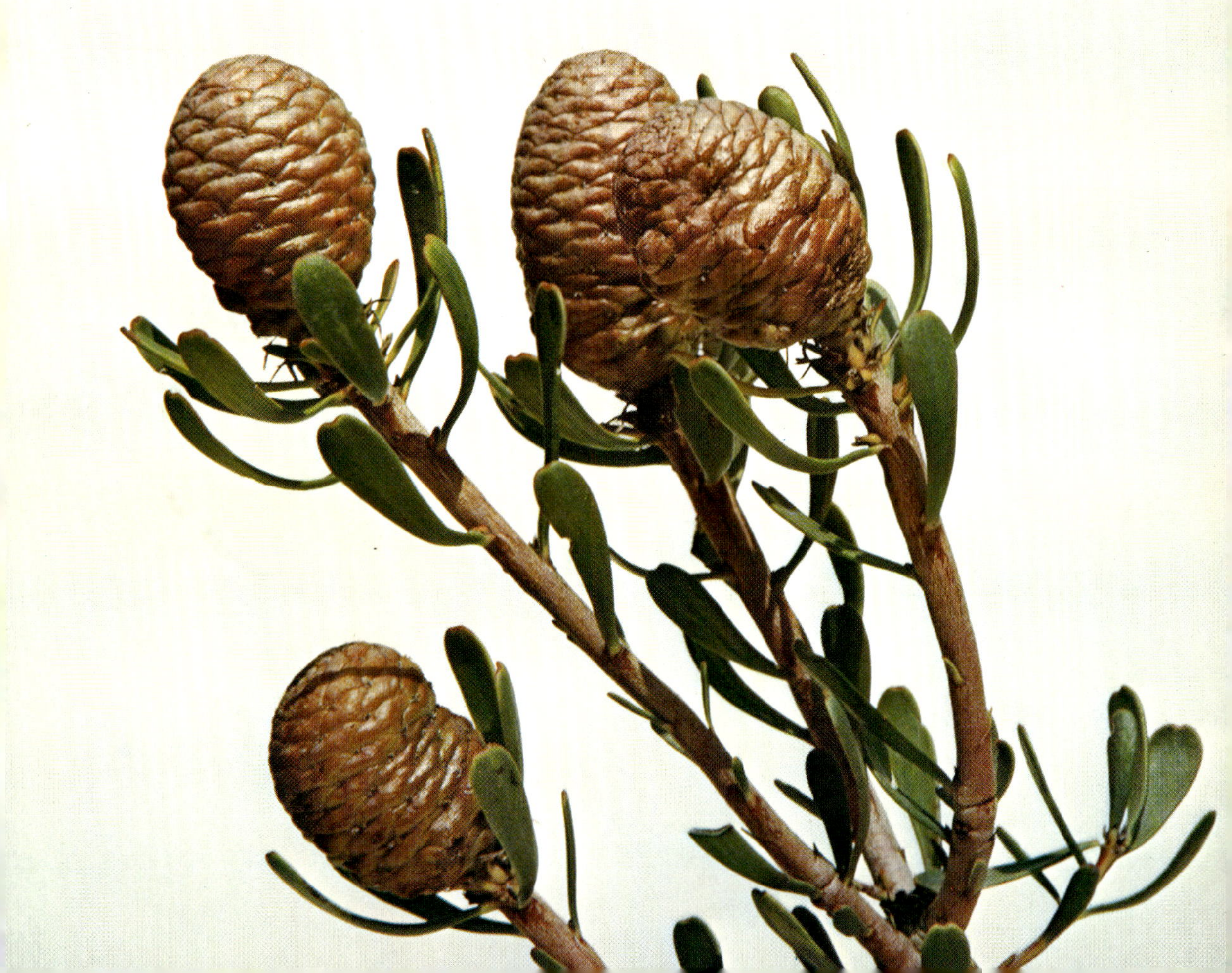

KNOBKERRIE BUSH

Leucadendron platyspermum

This 6-foot shrub is of value for its interesting knob-like cones, which become woody when dry and make fascinating everlasting material for the vase. The cone opens in spiralled, horizontal layers to release the flat, disc-like seeds that resemble butterflies. These lie on the "shelves" and sift out gently to fly away on the wind. The ridged, dark brown cone remains attractive for a long time and is much prized by florists.

The male flower, visible on the left-hand top photograph, is insignificant, being a fluffy head less than half an inch across. The foliage of the male flower becomes yellow near the top in the spring. The foliage of *L. platyspermum* is interesting as there are two types of leaves on the same bush (dimorphic foliage). The ordinary leaves are about two inches long and a quarter-inch wide, changing imperceptibly into soft, needle-like foliage near the tips of some of the branches. The male plant has more slender leaves. This species is easy to cultivate. It grows wild in the south-western Cape near Piquetberg, Caledon, Bredasdorp, Riversdale and Langeberg.

TOLBOS, TOP-BUSH

Leucadendron rubrum (L. plumosum)

This another species with a dried cone that is prized by those who arrange flowers. Both male and female flowers are illustrated opposite, below.

Again it is the female that is attractive. Its onion-shaped cone is tightly covered with overlapping bracts that are dark grey, tipped with brown. As it resembles a child's top, the bush has been given the common name of Tolbos ("tol" meaning a top). As the flower-head matures, it opens into a woody rosette of many pockets, each releasing a tiny pointed seed attached to a hairy parachute that wafts it to the ground. The fresh cones will dry out and release their seeds even in the vase, while the woody cone will last indefinitely.

The male plant is unlike that of most Leucadendrons. Each branch looks like a plume of gold made by hundreds of tiny flowers that appear in thick clusters near the tips of the branches and extend downwards. These flowers do not last long in the vase. They bloom in spring and early summer.

L. rubrum reaches a height of about 6 to 8 feet. It grows wild in sandy soil to altitudes of 3,000 feet and is common in the whole of the south-western Cape, extending eastwards to Uniondale.

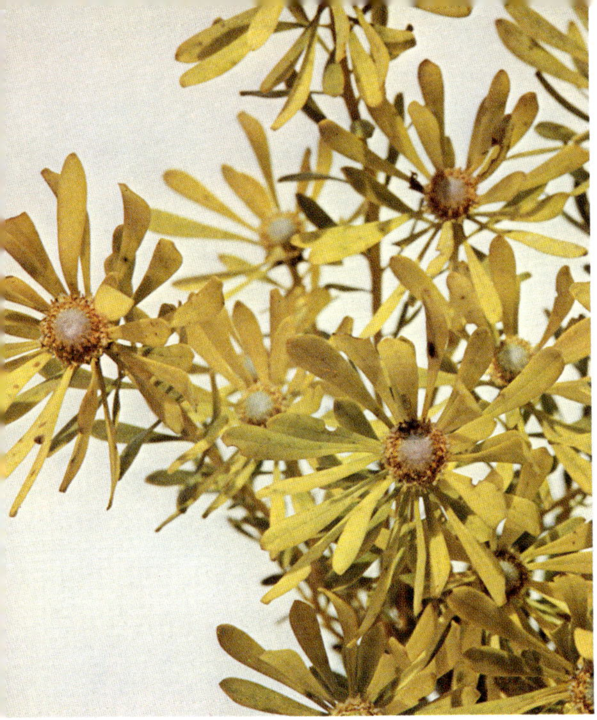
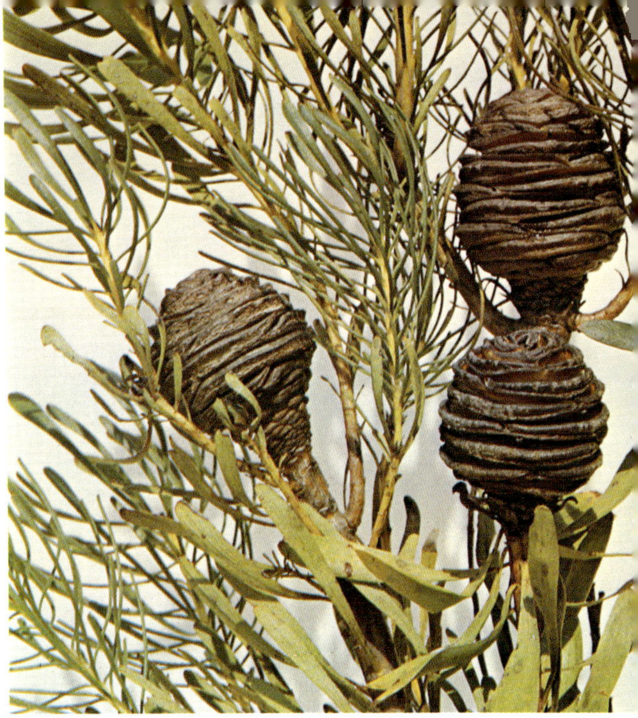
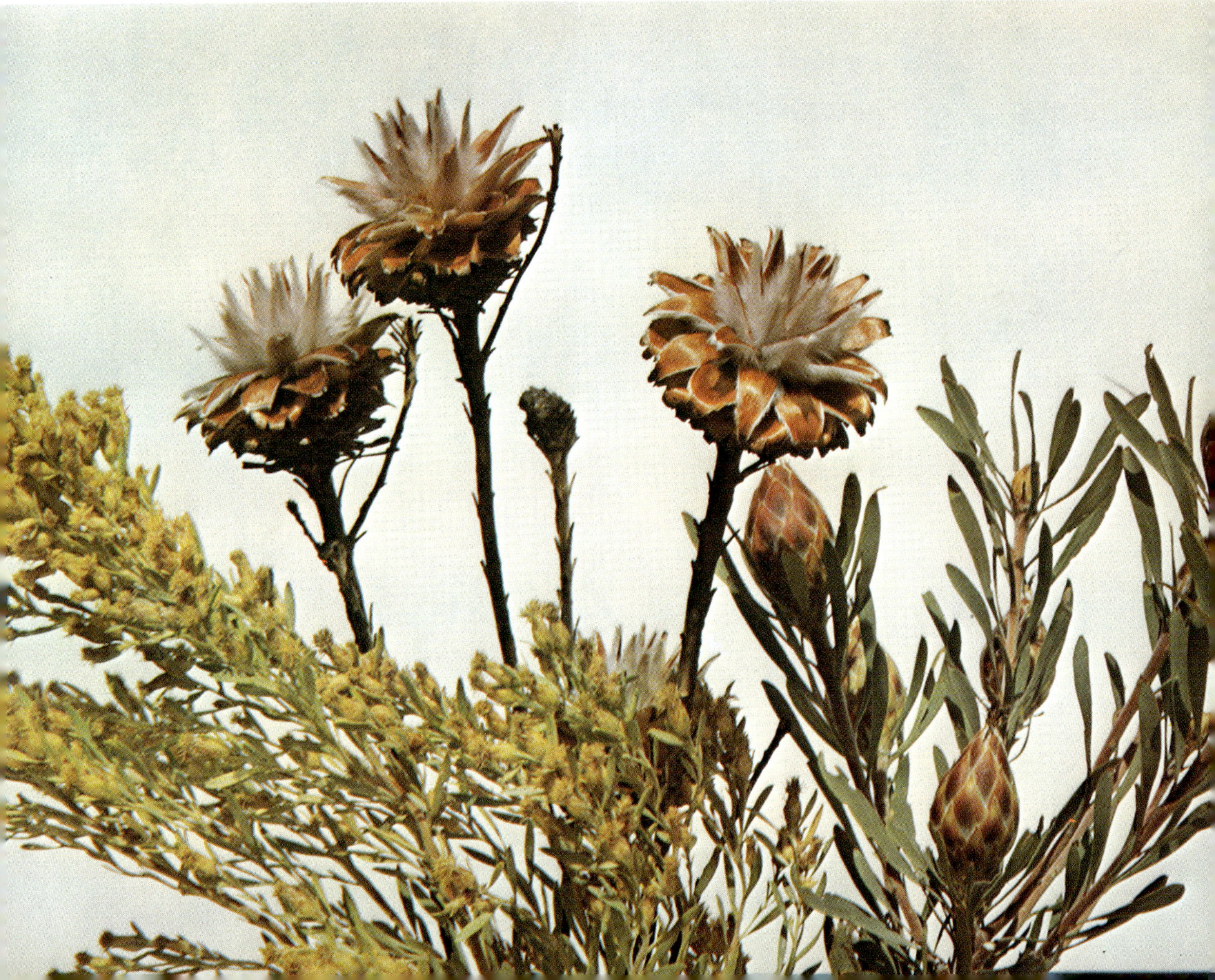

SILVER OR GREY LEAVES

Leucadendron spp.

In addition to the famous *Silver Tree* (*Leucadendron argenteum*), there are several smaller plants with grey or silver leaves, which may be grown for interest in a large garden. Many people specialise in plants with grey or silver foliage, which mixes well with white flowers, and the following are useful, easy to grow and available in the trade.

L. uliginosum. SILVER BUSH. This tall plant grows rapidly to a height of five feet, with many upright stems that often develop graceful bends that are useful for decorative purposes. The grey-green leaves are short and tapering, just over an inch in length, and covered with silvery hairs that glint in the sun. Short twigs last for weeks in the vase and mix well with white flowers. The male fluffy flowers and small female cones are really insignificant as this plant should be grown only for its silvery foliage.

This species is frost-resistant and will grow in wet or dry soil. It is sometimes found on the banks of streams in nature. It grows in limestone soil in nature and will grow in acid or alkaline soils in gardens. Full sunshine is best for the foliage. This species is common in the south-western Cape from the Peninsula to George. (See illustration opposite.)

L. pubescens (L. sericocephalum) is another tall shrub with silvery-grey leaves that are broader than those of *L. uliginosum.* It has silvery cones. This species grows wild in the Ceres and Clanwilliam areas.

L. album (L. aurantiacum) is an attractive 4-foot species with silvery leaves and cones. It has fine needle-like foliage and the female has silvery cones. (See opposite, below). The male has numerous yellow fluffy flowers in plumes like those of *L. rubrum.*

This is a frost-hardy species that will grow in alkaline as well as acid soil. It is useful in sea-coast gardens and grows easily inland. It is common at the Cape in the Outeniqua mountains and in the Albertinia district, where it grows close to the sea.

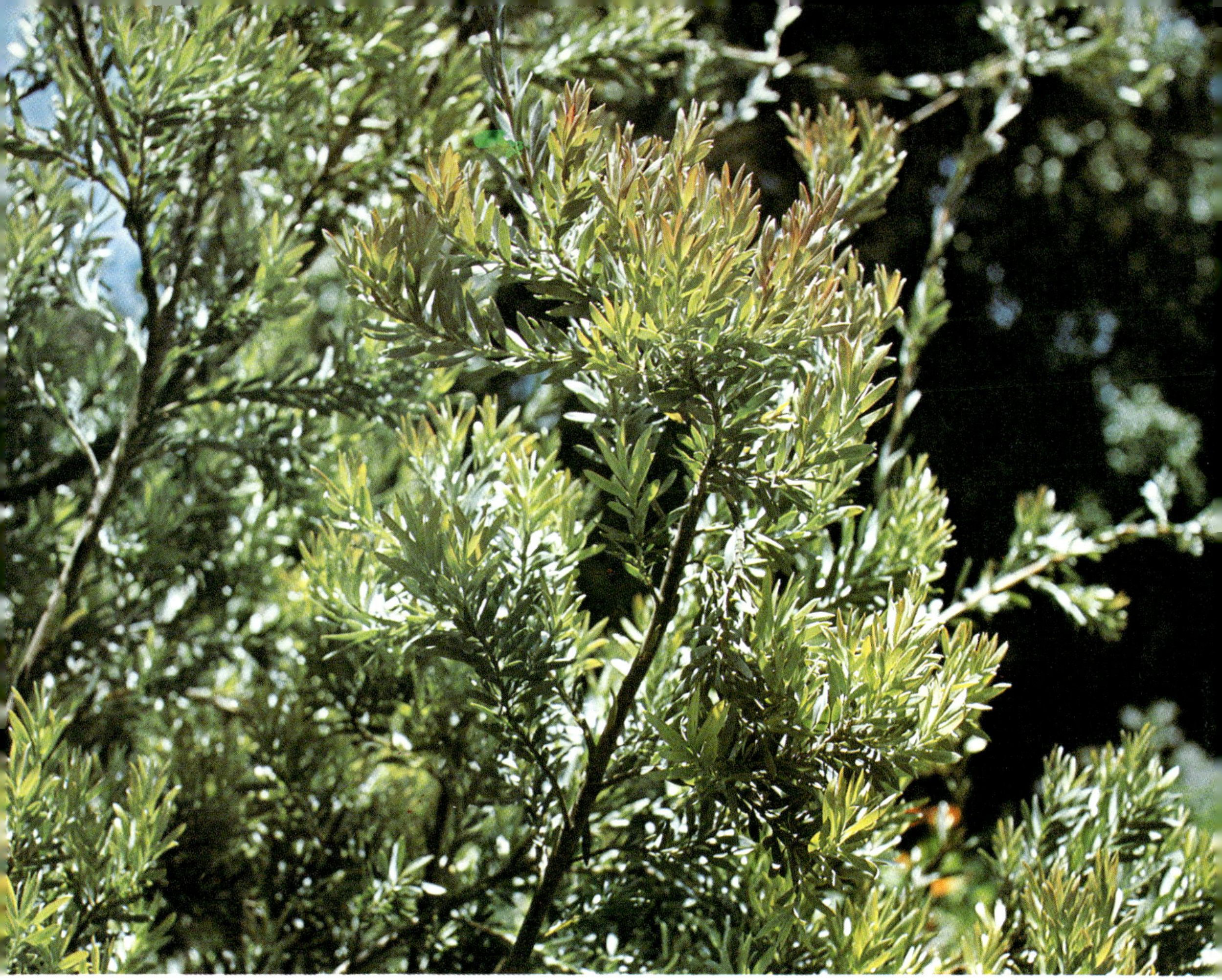

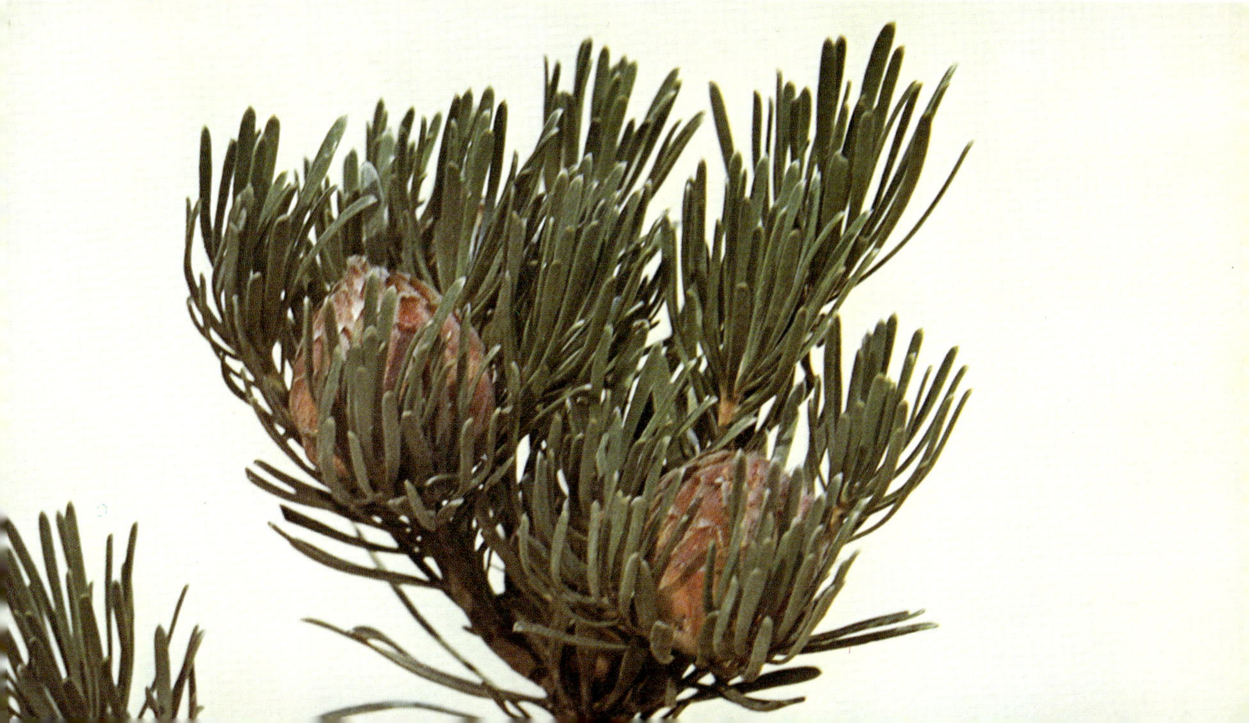

GALPIN'S LEUCADENDRON
Leucadendron galpinii

This evergreen shrub is available in the trade and grows almost without attention in the garden. It is a graceful, erect bush that grows to a height of five to six feet. Although it has no great beauty it is a useful evergreen for the larger garden.

The male flower, pictured here, consists of a small, fluffy, creamy pompon, measuring about half an inch across. These are massed on the twiggy branches in late spring. The female plant bears small pink cones. This is a common species in the south-western Cape. It should be watered during winter and given well-drained soil and sunshine. It is hardy to frost.

Several other Leucadendrons that grow easily and are available in the trade, could be grown in the larger garden, but should be a secondary choice when compared to the more showy species. These include the following:

L. comosum (L. aemulum.) This four to five foot species has large brown cones on the female plants. The male flowers are brown and fluffy—about one third of an inch across. The dense foliage is of two kinds (dimorphic), the lower leaves being thin, needle-like and short and the upper leaves about one-quarter of an inch wide. They are smooth and deep green. This species comes from the Cape mountains near Paarl, Swellendam and Port Elizabeth.

L. linifolium (L. tortum) has thin delicate green leaves and tiny cones. It grows to about 4 feet. (See opposite, below).

L. cinereum, four to five feet, has tiny green cones.

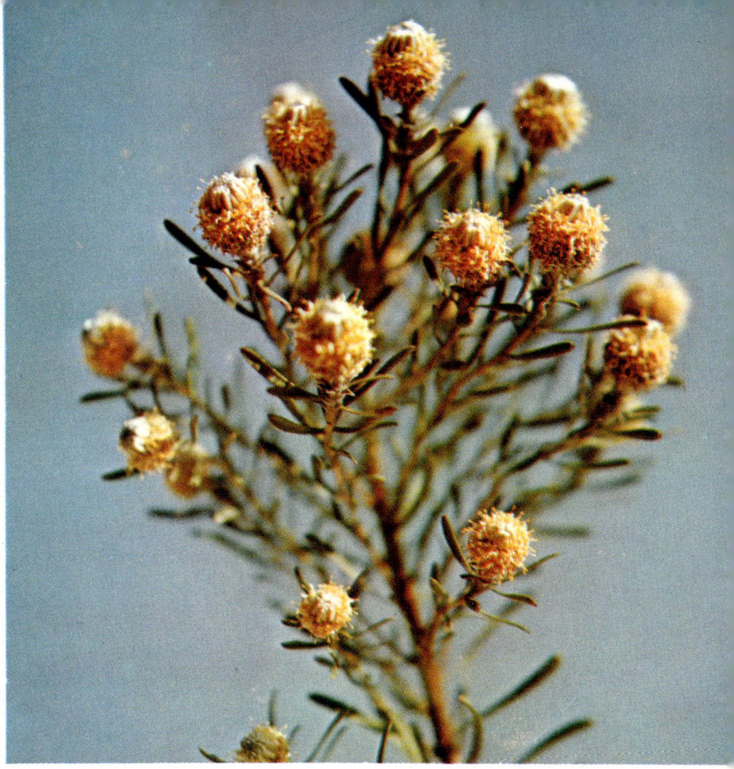

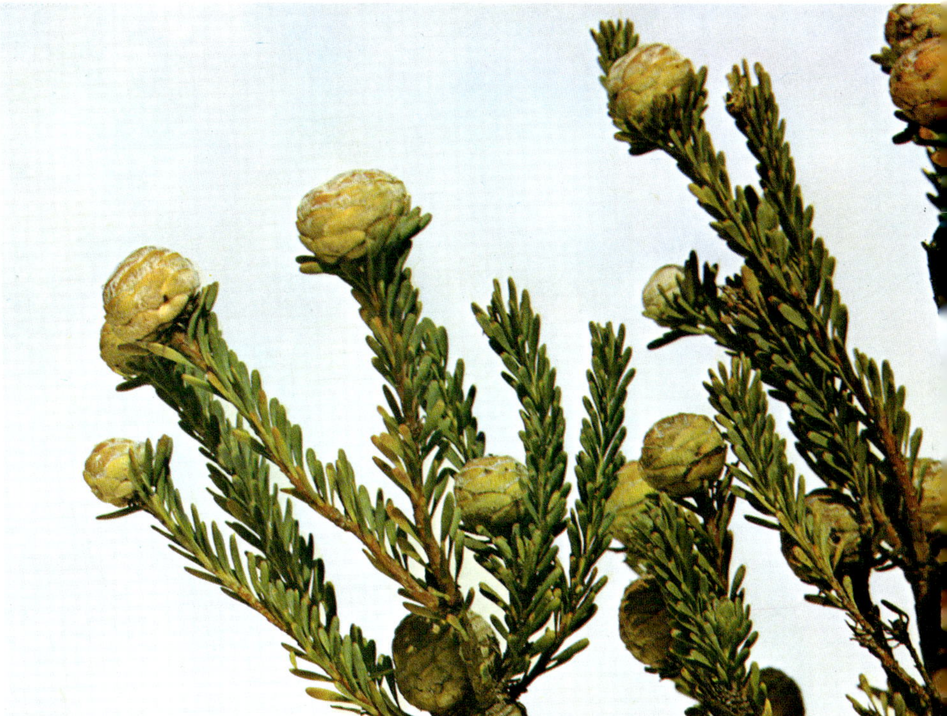

RED BOTTLEBRUSH, ROOISTOMPIE, SOLDAAT
Mimetes cucullatus (M. lyrigera)

The only species of *Mimetes* which is available in cultivation, this lovely plant is gay and colourful. If one drives from Bot River to Betty's Bay in the Cape, during spring and up to midsummer, one can see the rounded 4 foot bushes covered with hundreds of upright red and yellow flower-spikes, as trim as the soldiers of early days with their colourful uniforms.

The decorative effect of the foot-long bottlebrush is obtained by the topmost leaves changing in colour to yellow near the base and shading to bright red near the tips. The flowers themselves peep from between the leaves near the top of the stem, consisting of a tassel of silvery, hairy stamens and projecting smooth red styles, each hooded by a scarlet bract. The leathery green leaves overlap symmetrically up the stems of the entire bush, hiding them from view. Each measures about $1\frac{1}{2}$ inches in length and $\frac{1}{2}$ an inch wide.

Mimetes does not grow well in dry soil, needing peaty soil containing compost and plenty of moisture throughout the year. It should not be waterlogged, but a spongy soil will assist drainage. It enjoys humidity but the foliage should not be watered, especially during the late summer. It is an excellent plant for sea-coast gardens, growing in sand near the sea in nature, in the Caledon area and on the Cape Peninsula. It dislikes extreme cold, but can resist 15 degrees of frost and needs full sunshine.

The genus *Mimetes* consists of several rare plants, which are seldom seen in nature and seed of which is unobtainable. *M. argenteus* is the most silvery and *M. hirtus* the most magnificent, but several others bear mentioning. *M. hartogi* has similar flowers to *M. cucullatus* and was once thought to be a variety of it. It grows into a very large bush, sometimes 15 feet in height, with a distinct trunk. *M. hottentoticus* has colourful orange styles. *M. capitulatus* has rainbow colours and the flowers open slowly over a long period from May to September.

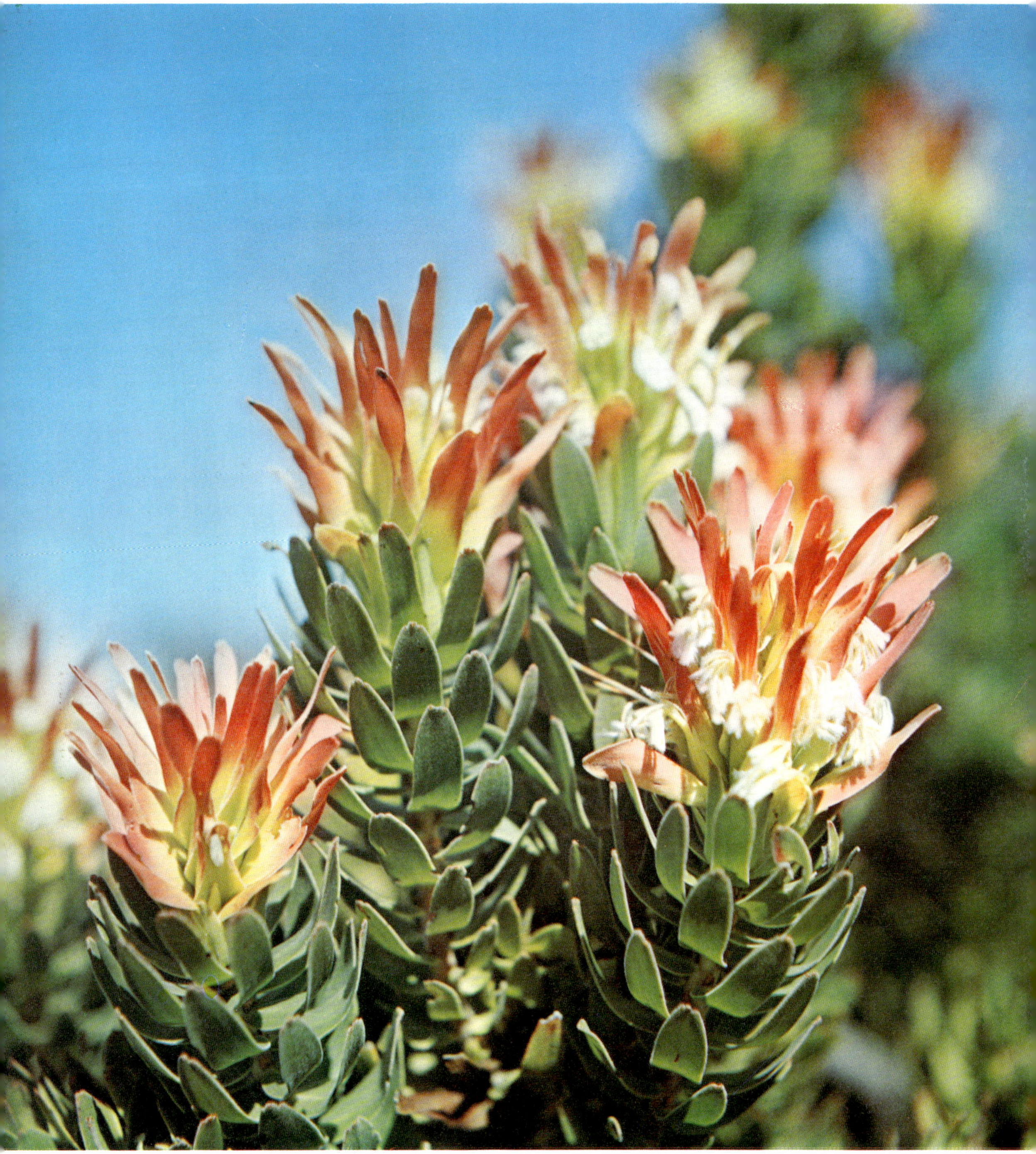

SILVER-LEAVED BOTTLEBRUSH
Mimetes argenteus

The silvery sheen on the upper part of the leaves of this plant makes it rival the Silver Tree. It has a more delicate appearance, however, for the leaves spread out gracefully from the reddish stems and those near the ends of the branches are shaded with old rose. The flowers themselves, which are grouped near the ends of the branches to form a bottlebrush spike, are not colourful. They peep out from between the leaves and are a combination of red and silver with a touch of yellow. The young leaves at the tip of the branch are very silvery. The bush grows to a height of about 4 feet and flowers in winter and spring.

This is a very rare species that is seldom seen in nature except by the mountain-climbing enthusiast. It grows in the south-western Cape mountains near the coast, where it is wreathed in cloud mists, or grows at the side of streams. It is not obtainable in the trade, nor is it likely to be for a long time. It could only be grown in cool, damp areas, preferably in mountain conditions where mist could provide moisture for this truly regal plant.

Other species with silvery foliage include *M. hottentoticus* and *M. integrus*. Neither of these is well-known or obtainable.

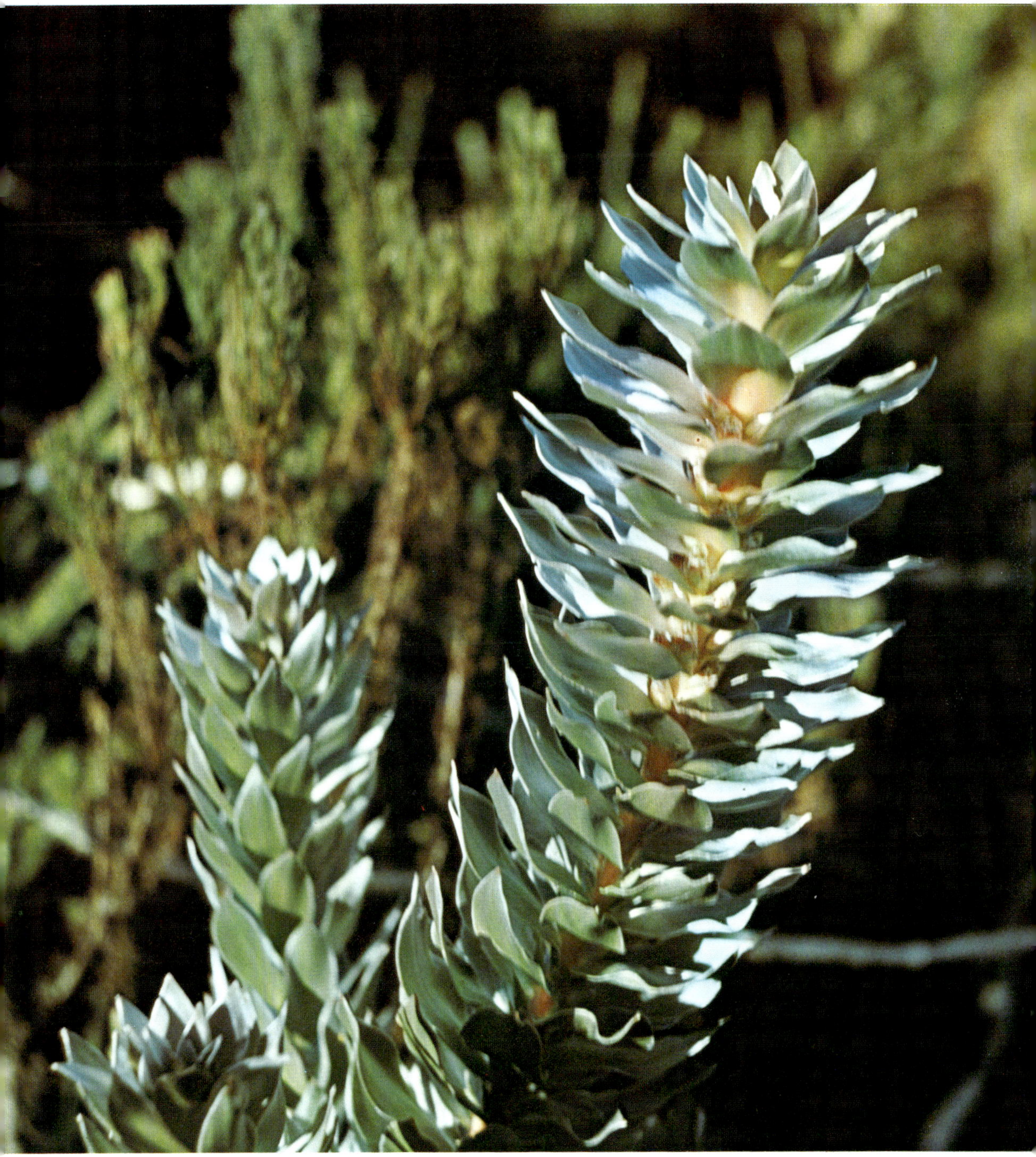

RED-AND-YELLOW BOTTLEBRUSH
Mimetes hirtus

If one should travel in September near Betty's Bay, Cape, one might be fortunate to see the magnificent flower spikes of *Mimetes hirtus* with its tall stems emerging through the bushes near the sea. The plants grow up to about 4 or sometimes 6 feet in height and are generally distinguishable only when they bloom. Otherwise they merge into the bracken and evergreen bushes that cover the hillsides with what is known as "fynbos" or "macchia". When they do bloom in the spring, they are most spectacular.

The individual flowers are massed towards the top of the stem, making a showy flower-spike that is long-lasting and colourful in the vase. The construction of the flowers is complicated to a layman, but the colour is provided by yellow scales that seem to form a thick tube. A long tuft of glowing red styles springs from the centre of this, while a fringe of hairy white sepals hangs downwards. At the very tip of the flower-spike, the topmost leaves form a little reddish-brown rosette that crowns the whole. The interest and colour of this Bottlebrush is difficult to convey, but it is a most desirable species. The leaves of *M. hirtus* are pointed and overlap upwards to cover the stem completely. They are plain green, but are softened by whitish hairs that also fringe the edges.

Bottlebrushes like plenty of moisture and will not progress in the garden unless they are watered well and regularly. They need good, peaty soil that should contain plenty of compost. They are not hardy to severe frost. *M. hirtus* is not freely obtainable, but would be well worth cultivating if it were offered in the trade. It grows wild at low altitudes in the Cape Peninsula as well as in the Caledon, Stellenbosch and Bredasdorp areas of the south-western Cape.

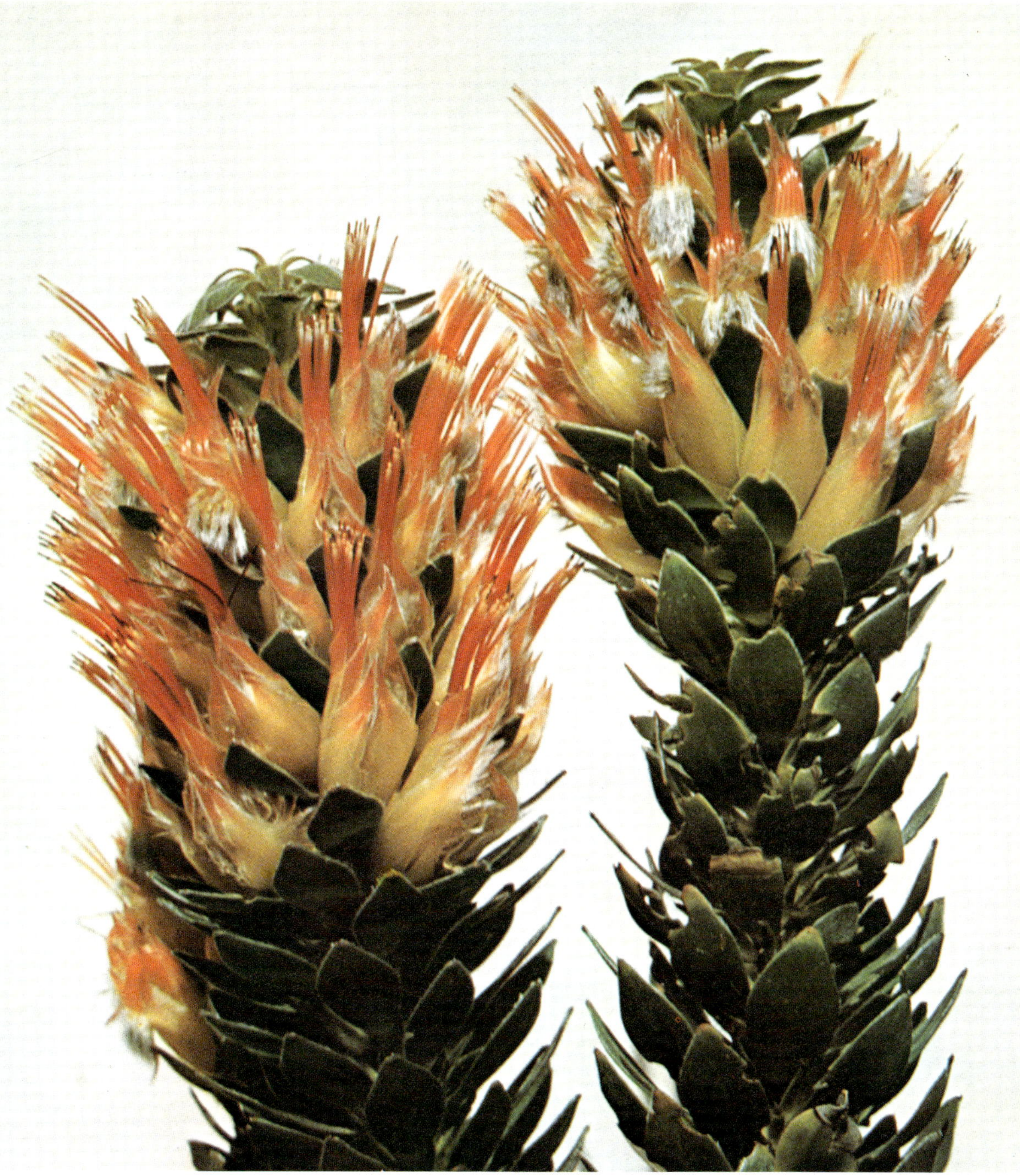

BLUSHING BRIDE, TROTS VAN FRANSCH HOEK
Serruria florida

Few people who see the Blushing Bride for the first time can resist its dainty charm. It is a conversation piece at flower shows and lovely enough to be used in a specimen vase. The flowers last for a long time, drying out almost imperceptibly, so that they can be used in dry arrangements. They last well out of water and make splendid corsages. They bloom in midwinter, which is a further attraction.

Buds start forming in May and open slowly till they reach perfection in July or August. The nodding, creamy white flowers are delicately flushed with pink and sometimes green. Like a *Protea*, *Serruria florida* owes its colouring to its bracts, which are broad, pointed and of a papery quality. Before the flower-head opens, the real flowers within are like silky pink hairs. They become erect as the flower-head ages and form a brownish tuft at the centre. Although flowers are generally still present on the bush in spring, they are no longer at their best. Each flower-head measures about 2 inches across. From 3 to 5 flowers are generally grouped near the ends of the branches. The leaves are finely divided into smooth, long needle-shaped segments with a soft, feathery appearance.

This slender shrub grows to a height of about 4 or 5 feet. It has only a few open branches, so that one should grow two or three in a group for effect. Pinch out the growing tip when the plant is about 6 inches high in order to encourage branching. Faded flowers should be removed with a good length of stem

Do not plant out the Blushing Bride while it is smaller than six to eight inches unless conditions are extremely favourable. It is not easy to cultivate. It must be given very well-drained soil containing compost in the summer rainfall area, for heavy summer rains frequently cause damping off. It needs regular water during winter and throughout the year in dry hot places. It is drought-resistant at the coast. Afternoon shade will prevent excessive drying out in hot areas. It grows in sandy, rocky soil on steep hillsides in nature. Shelter from frost is essential in cold climates. It likes fresh air, but will thrive in a north-facing position in a rockery, in front of other shrubs or even near a warm wall on the highveld.

Seed planted in March germinates in about three weeks. Seedlings should be grown in individual tins until large enough to be planted in the garden. Do not plant them out before the weather has warmed up in the spring in cold climates. Young plants may be set out in autumn in the winter rainfall area. The plants will flower in the second year and have a life span of about seven years.

The Blushing Bride was discovered by Thunberg in the Fransch Hoek Mountains and then lost for over 90 years until it was rediscovered by Professor Macowan in 1891. Later a guard was put over the existing plants by Kirstenbosch until the seeds ripened and plants were cultivated successfully at the National Botanic Gardens. Seed was then made available to growers and anyone can now buy plants of this rare treasure that was in danger of becoming extinct.

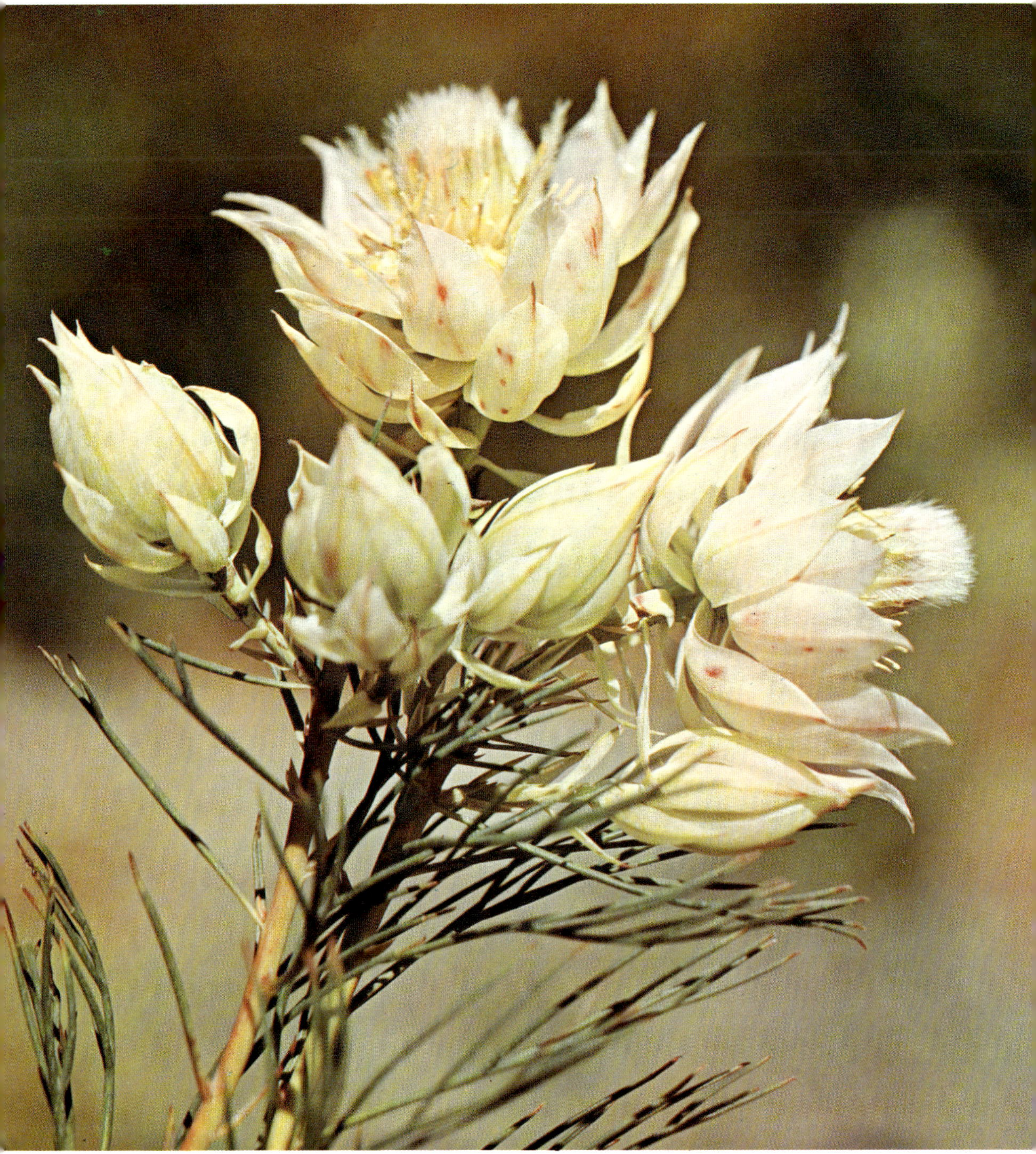

SILKY SERRURIA

Serruria barbigera

This is a small soft plant with soft, silky flowers. Each little head, about 1½ inches across, has tufts of pale pink flowers with silky hairs at the tips, so that they have a silvery-pink appearance. They are arranged in clusters at the tips of the branches, which are twiggy, so that several groups often appear as one large cluster. They bloom in spring.

The bush grows to a height of about two feet. It has the typical needle-shaped divided leaves of this genus, but this is a more branched and denser bush than the Blushing Bride. *S. barbigera* comes from the south-western Cape and should be cultivated like *S. florida*. It is available in the trade.

Several other dainty Serrurias are as follows:

S. burmanii, the Spider Bush or Spinnekopbos, is a three-foot shrub with very fine, dainty foliage and small clusters of tiny pink flowers. It is illustrated in *Wild Flowers of the Cape of Good Hope* by Rice and Compton. It grows wild in a large area of the south-western Cape, from Worcester to George.

S. knightii is now included under *S. burmanii*. (See opposite, below).

S. pinnata is a plant of botanical interest. It is a dwarf, creeping species with a small, silky pink flower-head at the tip of the branchlets. The leaves grow to one side of the stem as in *Protea acaulos*.

S. adscendens is another creeping Cape species, which is available in the trade. The spreading branches turn upwards at the end, bearing the finely divided leaves, while the clusters of tiny pink flowers nestle amongst the foliage like small pompons.

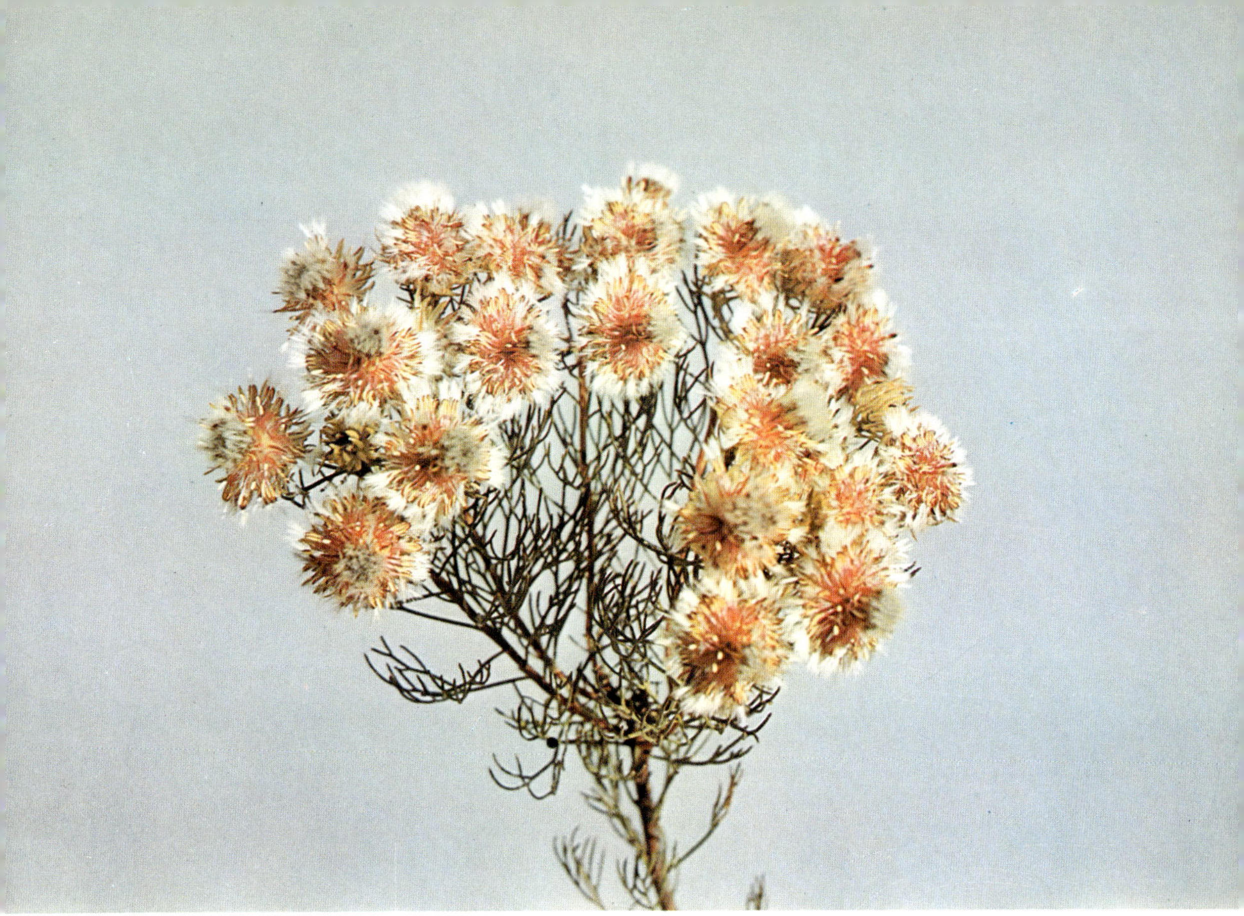

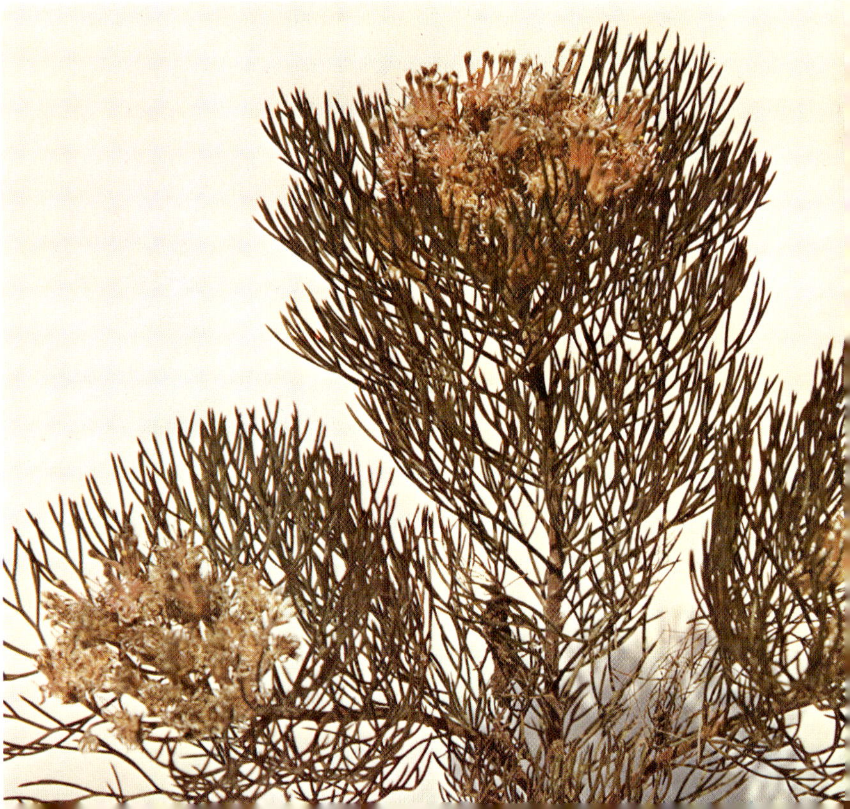

GREY SERRURIA

Serruria pedunculata (S. artemisiaefolia)

This low bush is interesting in the landscape because of its grey effect. The soft, finely-divided leaves are a dark grey-green and grow in tufts all over the plant, making it dense and compact. The small flower-heads, which measure about $1\frac{1}{2}$ inches across, are silvery grey. They resemble pompons which are made up of thin black threads edged with greyish-silver fluffy hairs. They are sometimes pinkish at the centre. The whole plant grows to a height of about two to three feet and spreads over the same area. The flowers appear freely all over the bush in the spring, and may flower in the first year if conditions are favourable.

S. pedunculata is easy to cultivate, needing sunshine and regular watering in winter. It can be transplanted into the garden when the bush is quite big. It will grow in ordinary gravelly soil that is fairly poor, but should be replaced after about seven years. Plants are available in the trade. This species grows wild in the south-western Cape, near Tulbagh, Ceres and Worcester.

S. aemula. This species has a tiny flower-head that measures an inch across. The outer pointed bracts are bright pink and the centre is filled with grey fluffy flowers. It blooms at the same time as *S. florida* and also grows wild at Fransch Hoek. An interesting hybrid with striped bracts was once produced by Frank Batchelor, using this species and *S. florida*. It is illustrated opposite, between its two parents. *S. aemula* also has finely cut, feathery leaves and should be cultivated in the same way as the Blushing Bride.

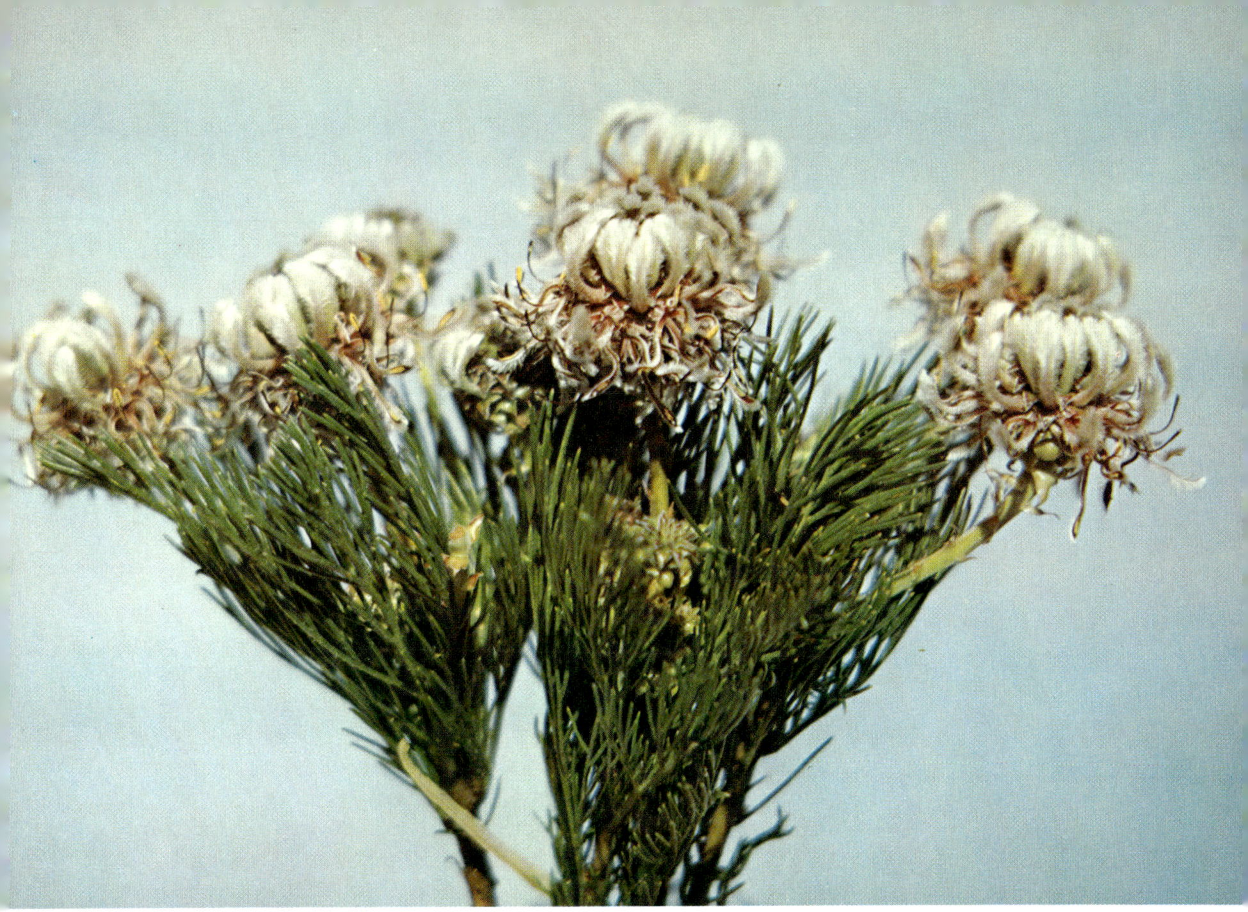

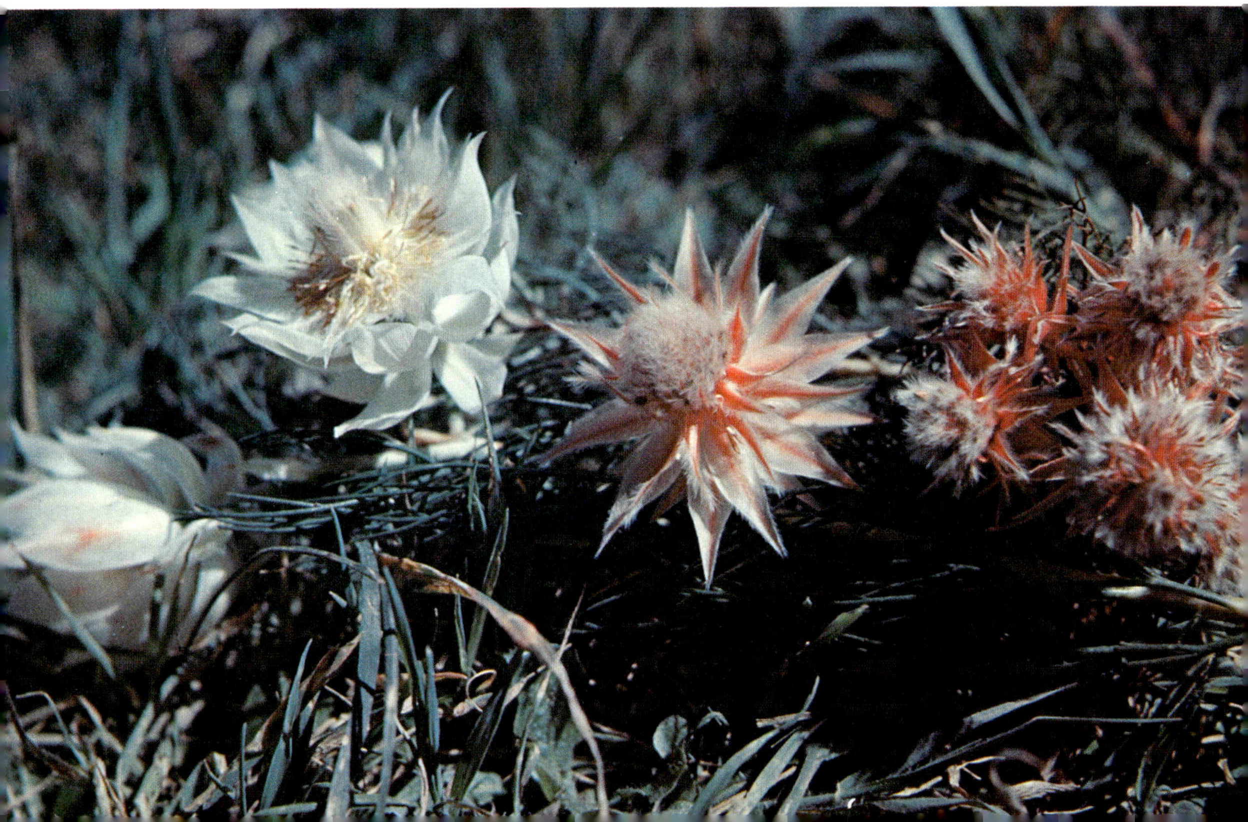

GREEN BOTTLEBRUSH

Paranomus reflexus

This bush, with its pale yellowish-green flowers that resemble a stiff bottlebrush, is an interesting evergreen shrub. The flowers last well in water and mix well with other members of the Protea family. The flower spikes are made up of slender tubes that reflex back as they age.

An interesting feature of this plant, which it shares with several species of *Leucadendron*, is that it has two different types of foliage on the same bush. The botanical term for this type of foliage is "dimorphic". The whole shrub, which grows up to about 4 or 5 feet in height, is covered with slender, feathery leaves. Shortly before it starts to flower, short foot-long branches grow out from the top, covered with short oval leaves, tapering towards the tip. These have long slender leaf-stalks. The appearance of these different leaves indicates that the plant is about to flower. The flower spikes begin to open in autumn and continue from May throughout winter and spring.

P. reflexus grows very easily in the garden in ordinary well-drained soil and is drought-resistant. It enjoys the shelter of other shrubs if it is grown in areas with cold winters. It grows wild in the Cape, being confined to the area from Humansdorp to Port Elizabeth, where rain falls sparsely, but at intervals through the year.

P. sceptrum-gustavianum is such a similar species that it can be confused very easily with *P. reflexus*. It has similar flowers, but these are shorter and do not reflex as they age. Also, the secondary leaves are wedge-shaped, being widest in the upper half. There is no leaf-stalk. This species is confined to the Caledon and Bredasdorp divisions of the south-western Cape.

P. spicatus (*P. crithmifolius*), Perdebos, is an attractive species because it has slender 4 to 5-inch spikes of bluish-pink flowers. There is also a form with red and grey flowers. These flower freely over the top surface of the bush in spring and early summer, creating a showy effect. The shrub grows to about $3\frac{1}{2}$ feet in height with a flat-topped shape. The leaves are feathery and greyish, and the plant does not develop different leaves at flowering time. *P. spicatus* grows wild in the south-western Cape on the sandstone hills near Clanwilliam, Caledon and Bredasdorp. It needs more moisture in the garden than *P. reflexus*.

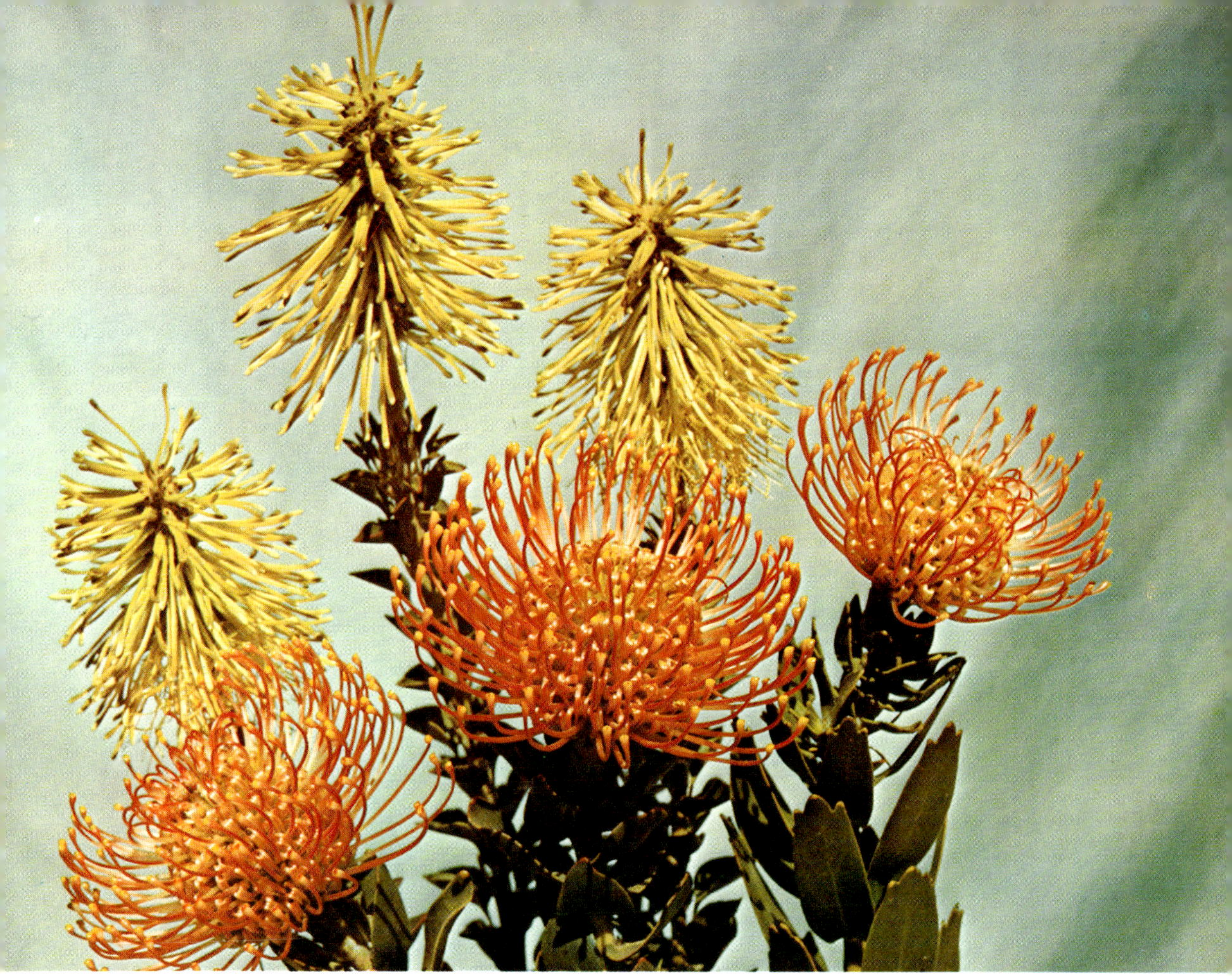
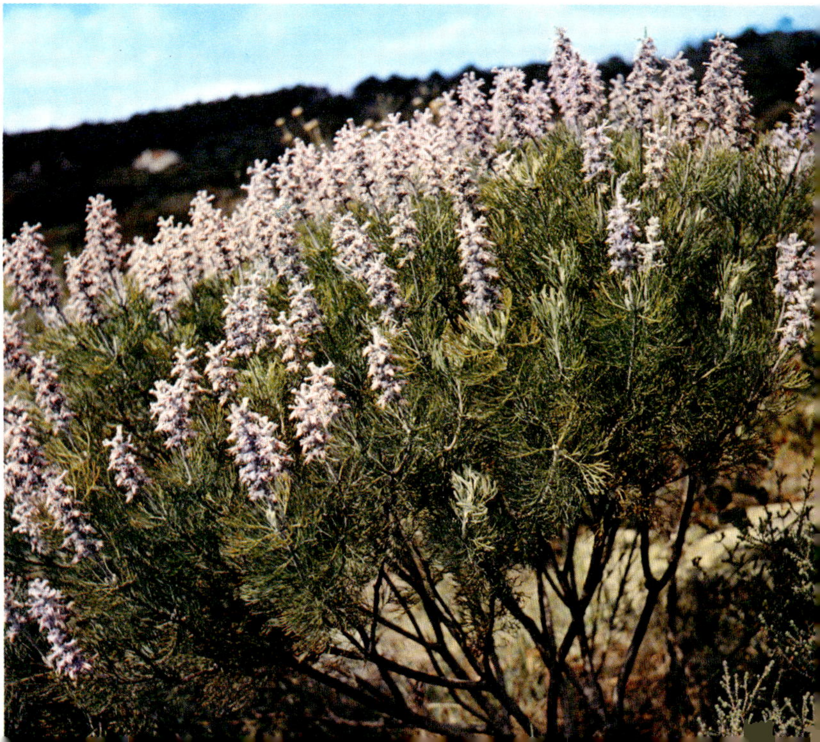

WILD ALMOND, WILDE AMANDEL, AMANDELHOUT

Brabejum stellatifolium

The Wild Almond will always be romantically linked with history, for this is the plant that was used by Jan Van Riebeeck in 1661 when he decided to plant a hedge to act as a boundary to the Colony and prevent the Hottentots from stealing his cattle. The remains of this historic hedge are still to be seen near the top of Kirstenbosch and it is now protected as a National Monument.

This is a very large spreading evergreen shrub that can grow up to 25 feet in favourable conditions. It has dense light green foliage and the pointed, 6-inch leaves are attractively notched. They are arranged in whorls at intervals around the stems, creating a starry effect. Small 2-inch spikes of fragrant white fluffy flowers appear in summer and the fruits ripen in autumn, from April to May. They look like almonds in shape and size, but have a brown velvety covering. They were described as "wild bitter almonds" by Van Riebeeck in his Journal and they have a bitter flavour. This is fortunate as they are poisonous, containing prussic acid. It is said that soaking the fruits for a few days removes the poison and that the nuts were roasted and ground for coffee in the early days.

Brabejum stellatifolium is the only species in this genus. It grows wild in the south-western Cape, being found near rivers and in moist places in the districts of Clanwilliam, Piquetberg, Ceres and Stellenbosch. It could be grown as a hedge in larger gardens at the Cape, but few people have space for it. Nevertheless, it makes an excellent dense evergreen to act as a screen. One must give it regular water during winter if it is grown in the summer-rainfall area, as well as a warm sheltered position.

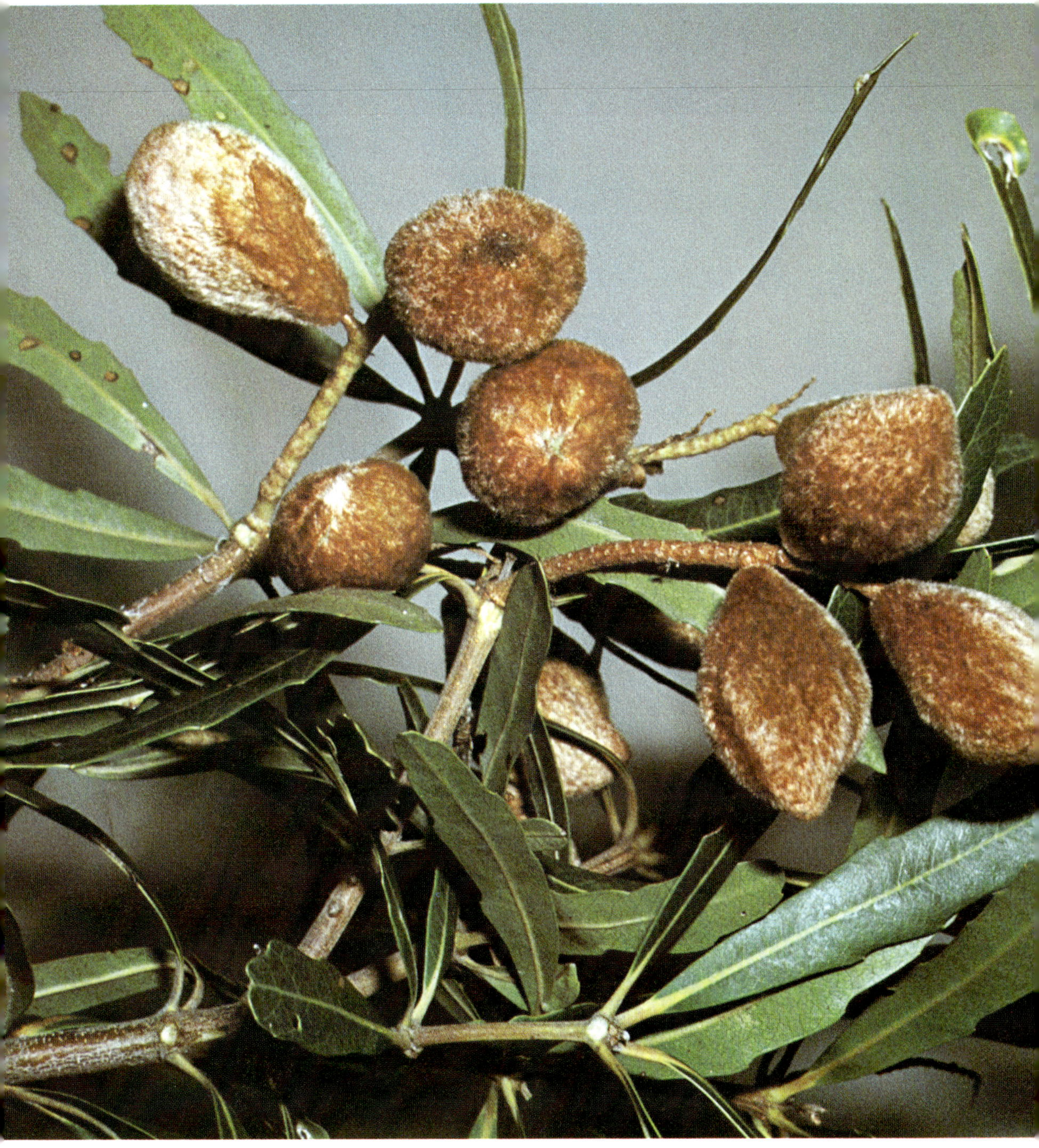

AULAX, FEATHER-DUSTERS

Aulax umbellata (A. cneorifolia)

This evergreen shrub grows to a height of 4 to 6 feet. It has male and female flowers on separate plants, in which respect it resembles the genus *Leucadendron*. The male flowers are bright golden yellow and arranged in tiny feathery plumes that are clustered together like a feather duster. When the shrubs bloom in spring or summer they create a colourful show that is of value in the landscape. They last well in water. The female plant produces small cone-shaped flower-heads which are inconspicuous, so that one should cultivate only the male. Both male and female are illustrated in the photograph. The narrow leaves of *A. umbellata* are slightly wider near the top, almost like a narrow spoon. They grow from about 1½ to 3 or 4 inches in length.

This shrub grows easily in gardens in a variety of soils, both acid and alkaline. It is reasonably hardy to frost and fairly drought-resistant, but requires water during winter. It grows wild in the south-western Cape near the sea-coast near Hermanus and Bredasdorp and on the mountains near Sir Lowry's Pass, Caledon, Riversdale and Mossel Bay.

Both this and the following species are available in the trade. There are only 3 species in the genus.

Aulax cancellata (A. pinifolia)

This can be distinguished from *A. umbellata* by its long, curved, pine-like leaves that grow up the stems at close intervals. The erect plant grows to a height of about 6 feet. The male plant is the more attractive to cultivate, having similar tufts of yellow flowers, grouped in feather-duster heads, to *A. umbellata*. The female plant has small 1½ inch-long cones at the tips of the branches, which may be yellow or green, shaded dark crimson. The female plant is generally smaller and more compact than the male. This species also grows wild in the south-western Cape and is easy to cultivate. It may be found near Ceres, Somerset West, Knysna and Uniondale.

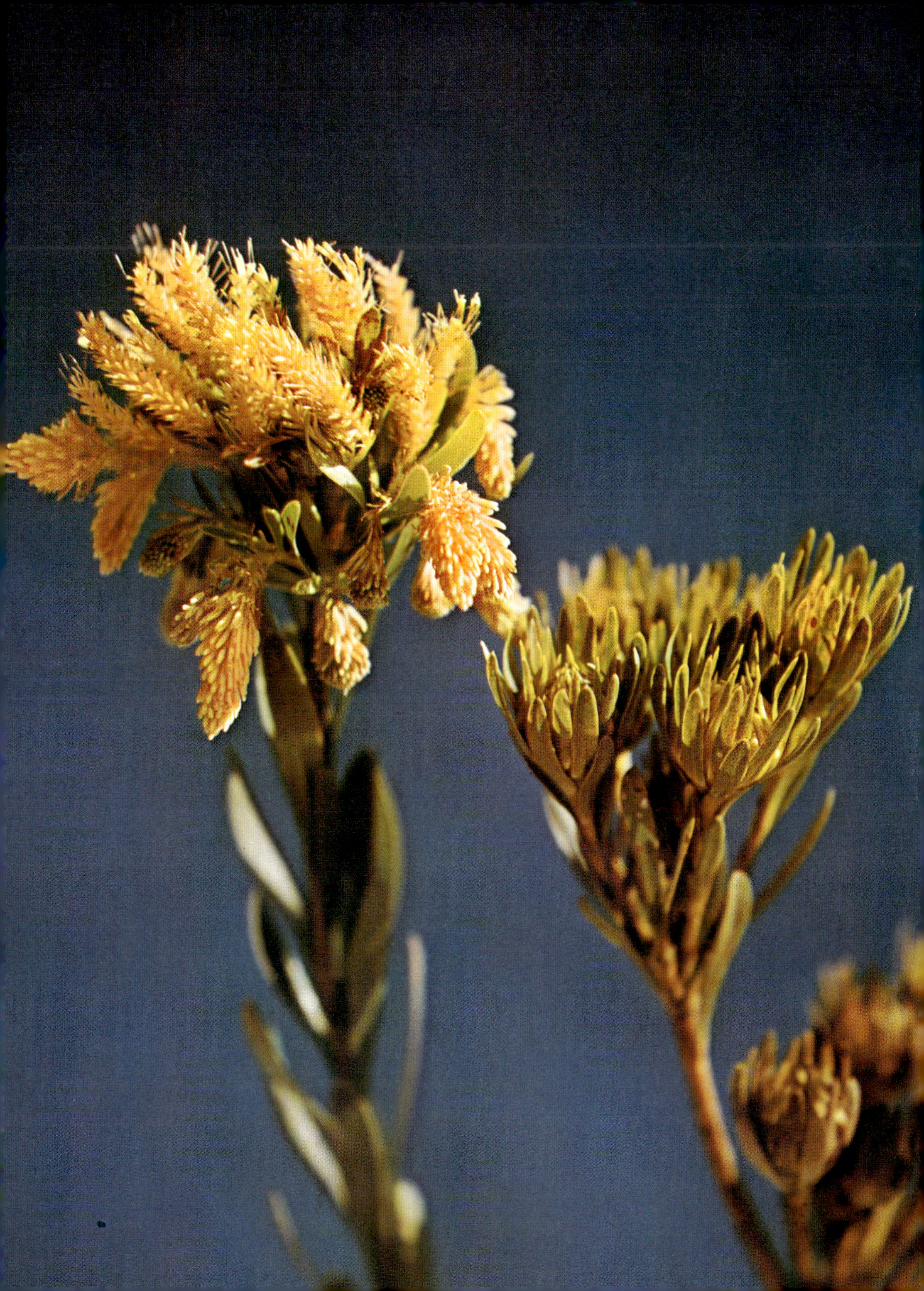

MARSH ROSE

Orothamnus zeyheri

Probably the rarest member of the Protea family, this unique and extraordinary flower is in danger of extinction. It is very restricted in the area in which it is found growing wild at the Cape and the few plants that exist are precious. Attempts to grow this plant have not met with complete success, although, at times, young plants have been reared and brought to the flowering stage, only to damp off and die later.

The flower is a curious contradiction in that the petal-like bracts, which are softly folded around one another like the petals of a rose, have a waxy, translucent texture, yet the outer surface is covered with soft white hairs that form a fringe at the tips. The flower is about 3 inches long and ultimately becomes a glowing pomegranate red. There are several flowers grouped at the tip of the stem, which is thickly covered with overlapping dark green leaves. These are oval, pointed and leathery in texture, with a light covering of soft white hairs which also edge the margins.

The Marsh Rose does not become a bushy plant, but has only a few long branches, some plants consisting of only one stalk. This grows to a height of 6 to 9 feet and the lower half is generally bare of foliage. If the flower is picked with a long stem, the plant dies, and the flower pickers of the past have done much to exterminate this flower, which is now protected by law.

The flowers are at their best in October and November, from spring to midsummer. They grow in a small area in the southern Hottentots Holland Mountains in the Caledon Division of the Cape, seen only by enthusiastic mountain climbers. This species, the only member of the genus, was lost to botanical science for many years and only rediscovered in recent years. The mountain slopes on which it grows are well-drained, consisting of peaty spongy soil through which moisture seeps slowly. It does not like marshy conditions, as the name suggests, and will damp off if the soil is too wet or the plant overwatered, particularly during summer. Seed will germinate in anything up to 8 months, and plants have been grown to 15 inches in a year, but this is a plant for the collector. The Marsh Rose has defied most attempts to bring it to maturity and it is thought that the cool cloud mists that bathe it in nature are essential to its well-being. Grafted specimens on *Leucospermum* rootstock have been produced by the Cape Department of Nature Conservation since 1976.

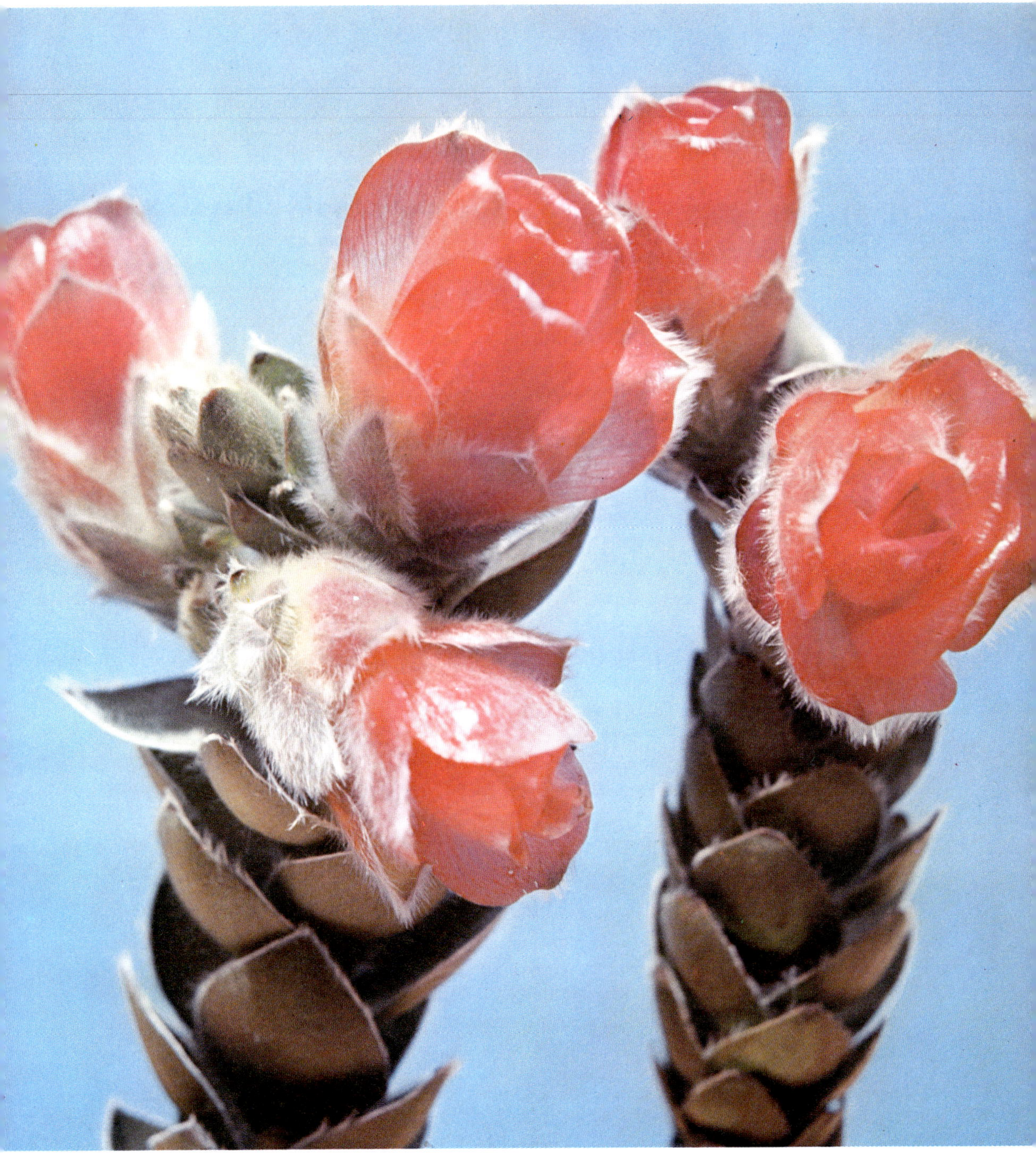

DIASTELLA

Diastella divaricata (D. serpyllifolia)

Surely the tiniest-flowered member of the Protea family, the small flower-heads are scarcely half an inch across, like miniature rose-coloured powderpuffs. They appear at the ends of twiggy branchlets which are covered all along their length with small oval leaves, about a quarter-inch long. This dainty plant creeps on the ground and reaches a height of about eighteen inches, so that it is by no means conspicuous, but a plant for the collector. It has a dainty charm that would appeal to the lover of small plants. The flowers last for a long time in water and the branches are graceful. It blooms almost throughout the year, especially in winter and spring. (See opposite, above).

D. divaricata is available in the trade and grows easily in gardens. It needs a well-drained soil containing sand and compost, but must be given regular watering, especially in winter. It grows on sandy flats in nature and should be grown in a rockery in the garden, both for the sake of its appearance and for successful cultivation. In areas with cold winters it should be given a sunny position where it is sheltered from frost. This species is common in the Paarl and Caledon districts of the south-western Cape, as well as on the Peninsula.

This is one of 4 species of *Diastella* which all come from the S.W. Cape, but none occurs further north than Tulbagh.

D. ericacifolia has finer, more compact, needle-like foliage, which is attractive. This plant creeps flat on the surface and should make a good plant for the rockery, when it becomes available in the trade.

Other genera with tiny flowers include *Spatalla,* with tiny spikes of flowers, and *Sorocephalus* with small heads. *Spatallopsis* is a genus that is now included with *Spatalla*. None of these is as well-known as *Diastella divaricata,* but it is possible that they will be grown when they are made available to collectors of the future. They are all to be found at the south-western Cape, where they form part of the "fynbos" or macchia that covers the hillsides with evergreen vegetation. The photograph depicts *Spatalla curvifolia*, a common species in the Caledon and Bredasdorp districts.

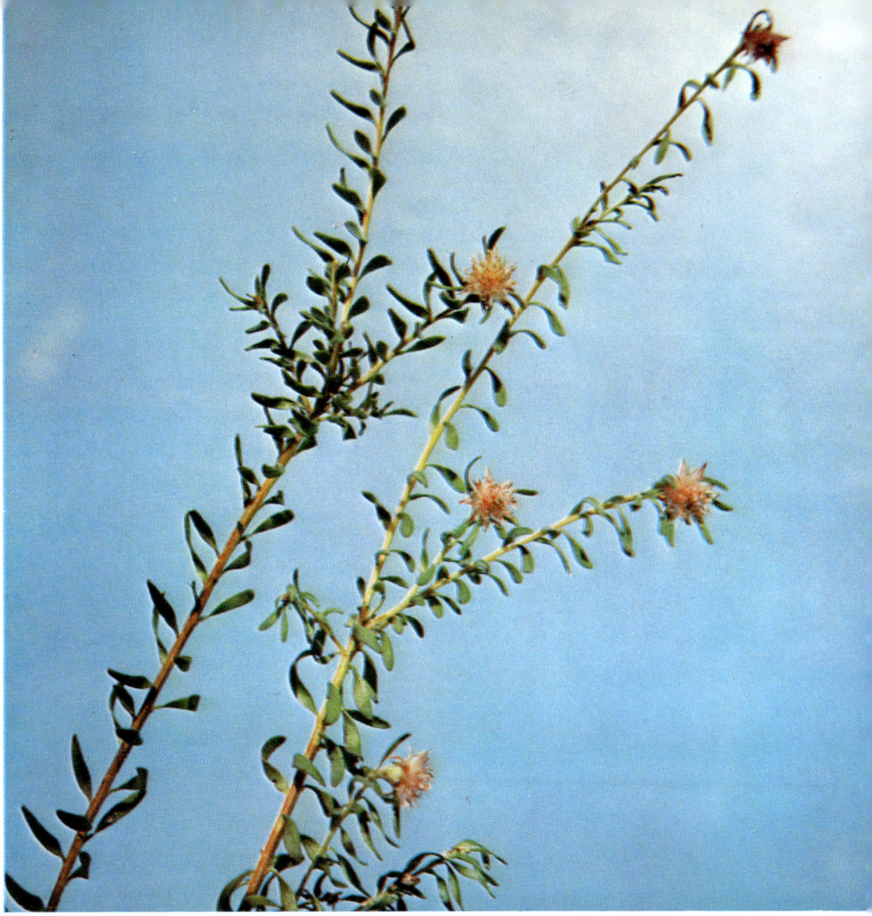

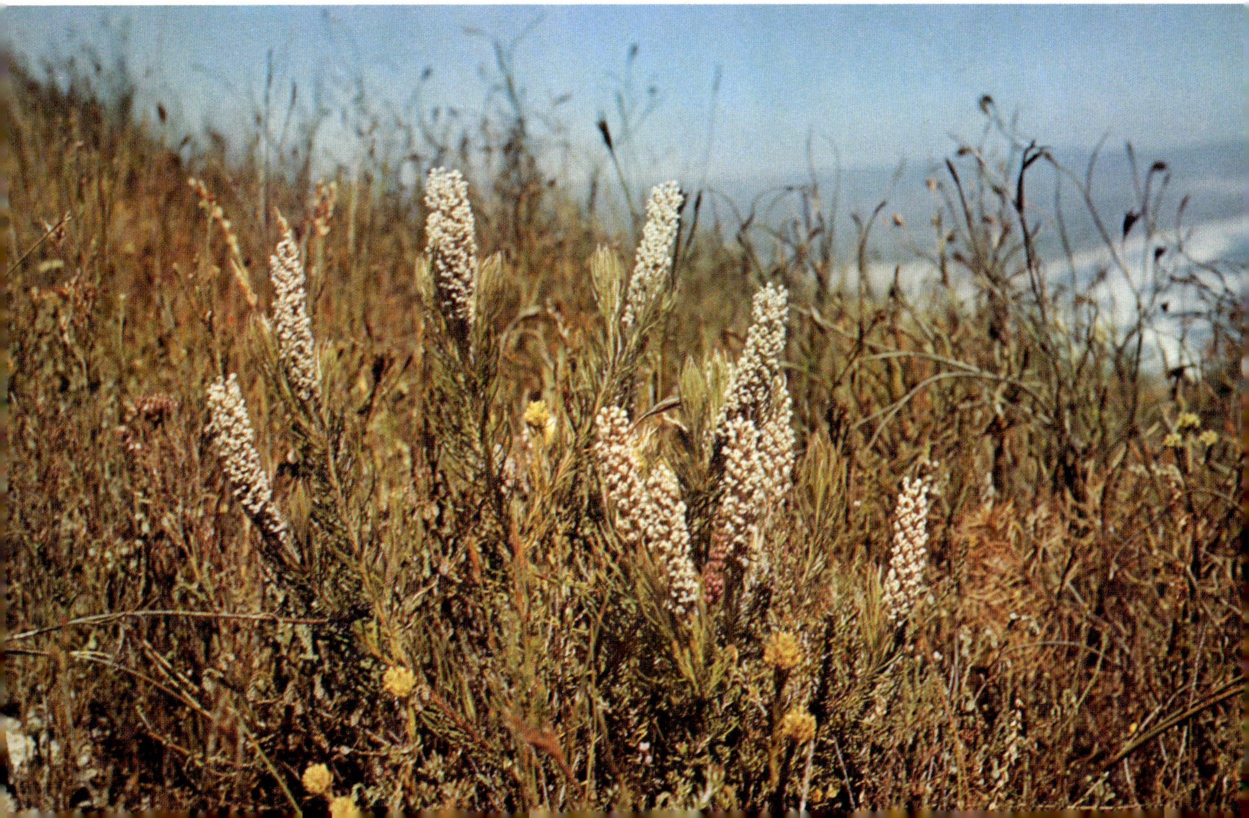

TRANSVAAL OR RED BOEKENHOUT
Faurea saligna

Unless one is told that this tree is a member of the Protea family, it is difficult to recognise it as such. It grows to a height of about 30 feet, although specimens of up to 60 feet have been seen. The flowers appear in spring, around September, together with the new leaves. They hang down in catkins, like pale, pinkish-mauve, silky tassels, about 4 inches in length, and have a scent of honey. *Faurea* differs from Pincushions and Proteas in that it has its flowers grouped in a long spike instead of a dense head. The long narrow leaves resemble those of the Eucalyptus. They develop reddish tints before dropping, but the tree is evergreen.

This graceful tree grows in the warmer areas of the Transvaal bushveld and lowveld, as well as in Natal, Zululand, Botswana, Zimbabwe and northwards into tropical Africa. It grows on stony hillsides and in sour, sandy veld. It is tender and could only be grown in areas with mild winters, but is not generally cultivated. It receives good summer rains in nature, but has a dry period of 6 months during winter and spring. *F. saligna* is the most widespread of the genus and *Faurea* is the only genus of the Protea family that is not found in the winter-rainfall area of the south-western Cape. *F. speciosa*, with much broader leaves, grows wild from Knysna to the north-eastern Transvaal. There are about 18 species of *Faurea* in Africa and one in Madagascar.

The wood of *Faurea* was likened by the early colonists to that of the European Beech (Boekenhout) and is attractively grained. It was used for furniture as well as wagonwheels and even as telephone poles. All species of *Faurea* are now protected trees in South Africa and may not be destroyed for economic use.

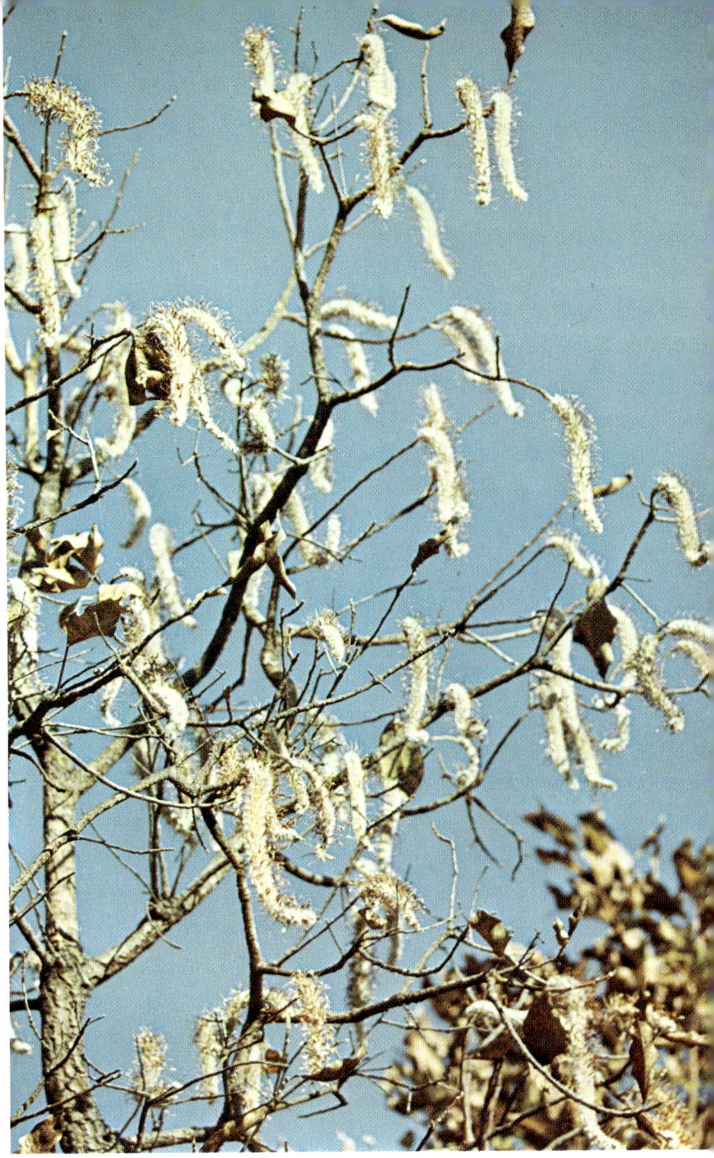

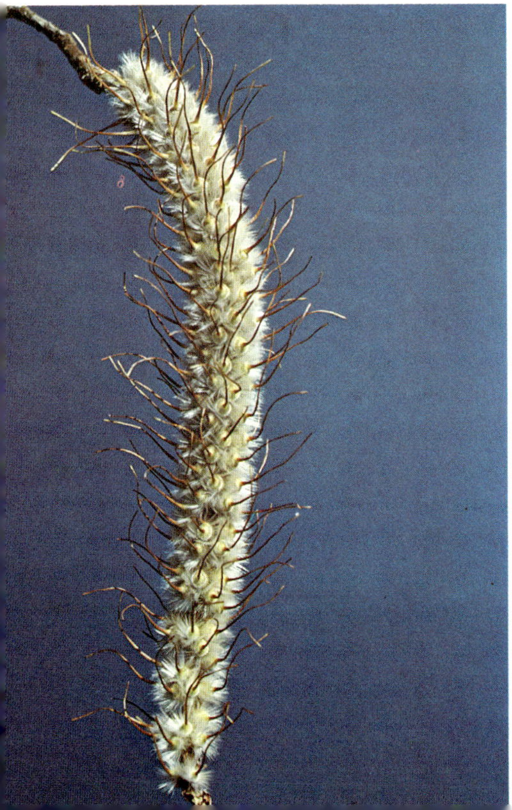

EXOTIC MEMBERS OF THE PROTEA FAMILY

Several attractive members of the Protea family which have been introduced from abroad, particularly from Australia, have been grown in Southern Africa.

Some grow with ease, while others present more difficulty. It is interesting to note these plants here, if only to present a more complete picture of the fascinating Protea family.

Grevillea (Australia, New Caledonia)

Several species of this attractive group have been grown for many years in Southern Africa, requiring ordinary garden soil and very little attention. The tall *Silky Oak* (*Grevillea robusta*) is a familiar tree in both street and garden. It flowers in October, when its golden panicles generally act as a foil to the Jacaranda, which blooms at the same time.

Scarlet Grevillea (*G. banksii*) has long been a favourite small tree in mild areas, blooming nearly all through the year, but it may also be grown on the highveld in sheltered situations.

Rosemary Grevillea (*G. rosmarinifolia*) is a six-foot shrub that has recently become popular in gardens because of its needle-shaped foliage and curled red flowers.

Telopea (Australia)

The *Waratah* (*T. speciosissima*), pictured overleaf, has a magnificent deep crimson flower-head that is reminiscent of a Pincushion and a Protea. It is grown at Kirstenbosch, despite its Australian origin, where it excites great admiration from visitors. Plants are available from nurserymen. It grows fairly slowly and needs protection from frost on the highveld.

Stenocarpus (Australia, New Caledonia)

The *Firewheel Tree* (*S. sinuatus*) has been grown in Durban for many years, but has only been made generally available to gardeners in recent years. Although the spectacular scarlet wheel-like flowers do not appear for about ten years, the beautiful glossy foliage makes this small tree worth growing.

Banksia (Australia)

Several species, notably *B. ericifolia.* have been cultivated here. It has needle-like leaves and burnt-orange flower-spikes. Scarlet *B. coccinea* and others are becoming popular.

Macadamia (Australia)

The delicious Queensland or Macadamia Nut (*M. ternifolia*) is now being grown as a commercial crop in Natal. It is extensively cultivated in Hawaii. This is a decorative tree for mild districts, with a round nut that has a very hard shell.

Hakea (Australia)

Several species are grown in Southern Africa, especially *H. saligna*, which makes a pleasant hedge plant. *H. sericea* (*H. acicularis*) and others, however, have become a menace to the indigenous vegetation at the Cape, for they grow so easily that they crowd out the local growth. This "alien vegetation" is being eradicated with great difficulty by Kirstenbosch, the Department of Nature Conservation and other protectors of our flora.

Embothrium (Chile)

The *Chilean Fire-Tree* (*E. coccineum*) is an evergreen shrub with scarlet Grevillea-like flowers in early summer. It has been grown in a garden in Kloof, Natal, and deserves to become more popular.

PLANTS FOR THE FUTURE

The following members of the Protea family have been cultivated in Australia, where they are indigenous. They should prove interesting plants for collectors if they are introduced to this country.

Dryandra (Australia)

Closely related to *Banksia*, these showy shrubs have large, Protea-like flower-heads, surrounded by rosettes of leaves. They are not easy to grow, nor is it easy to obtain seed.

Conospermum (Australia)

These shrubs are called *Smoke-Bush* because of their fine grey-woolly flowers that give the appearance of bush-fire smoke from a distance.

Isopogon (Australia)

Small shrubs, known as *Cone-Bushes*, these have showy cone-like purple, pink, yellow or grey flowers and prickly, pine-like leaves.

Persoonia (Australia, New Zealand)

These trees are commonly called *Geebung* and are valued for their berries. *P. pinifolia* has yellow flowers and is most attractive.

Petrophila (Australia)

These shrubs have spikes of grey-pink flowers that put one in mind of *Paranomus spicatus*.

Xylomelum (Australia)

A small group of trees and shrubs, these are called *Woody-Pear* because of their strange woody fruits that resemble a pear. They prefer mild conditions.

THE WARATAH
Telopea speciosissima

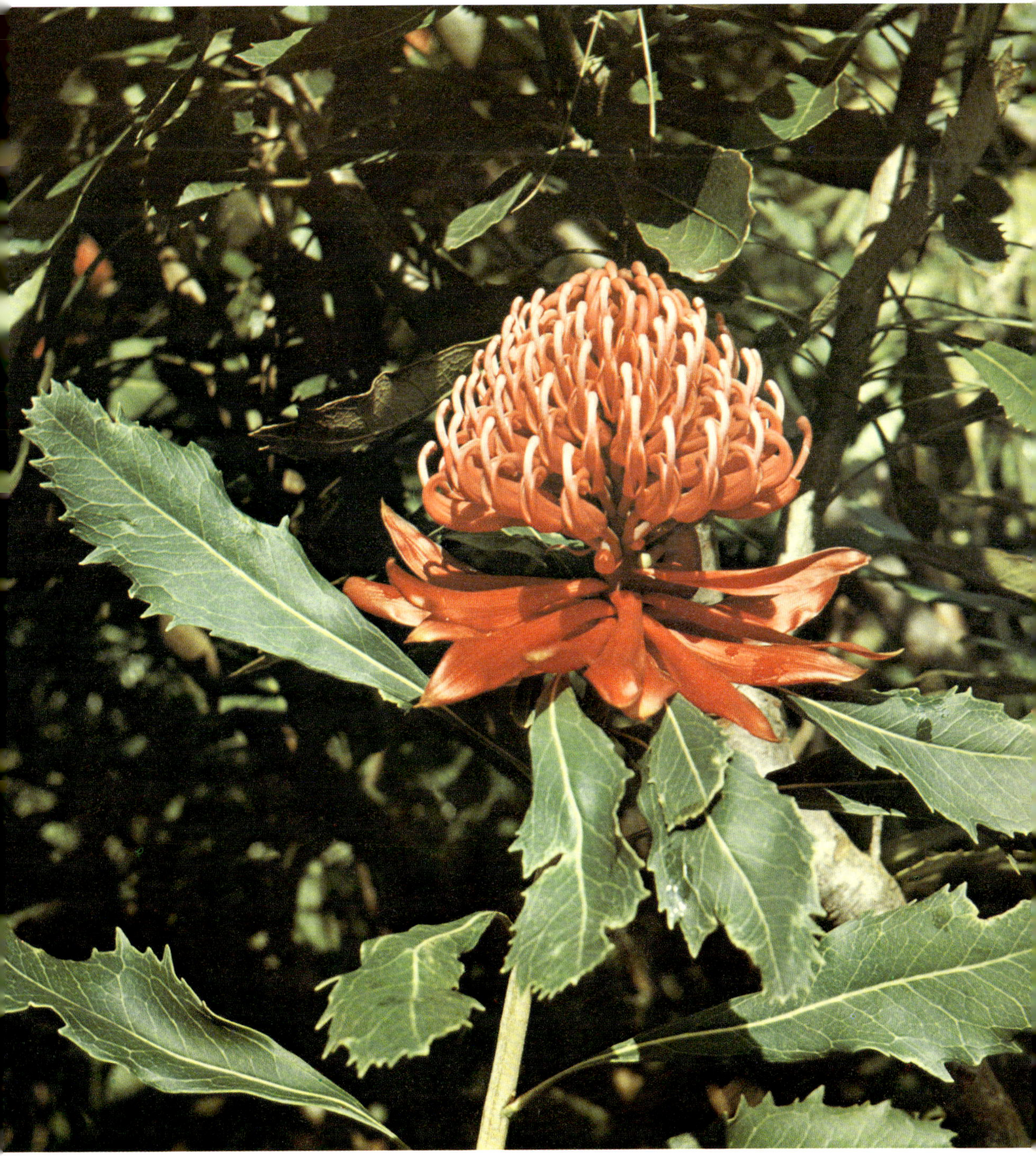

. . . sugarbirds,

that dip into the gleaming cup,

while swaying gently in the wind.

The Cape Sugarbird (*Promerops cafer*), perched on a flower-head of *Protea aurea*, is feeding on nectar and the insects which are found in these flowers. This male bird has longer tail feathers than the female. The latter builds its cup-shaped nests in Protea bushes and lines them with dry Protea wool.

The Sugarbirds belong to the only family of birds peculiar to Southern Africa and there are only two species. The Cape Sugarbird lives only in the southern Cape, while Gurney's Sugarbird occurs in the eastern Cape, Natal and Zimbabwe. Sugarbirds have long, curved bills which are ideal for probing into Protea flowers. Sunbirds have similar curved bills and the Orangebreasted Sunbird (*Nectarinia violacea*), in particular, feeds on the nectar of Proteas, as well as of other flowers, and on insects. (See next page.)

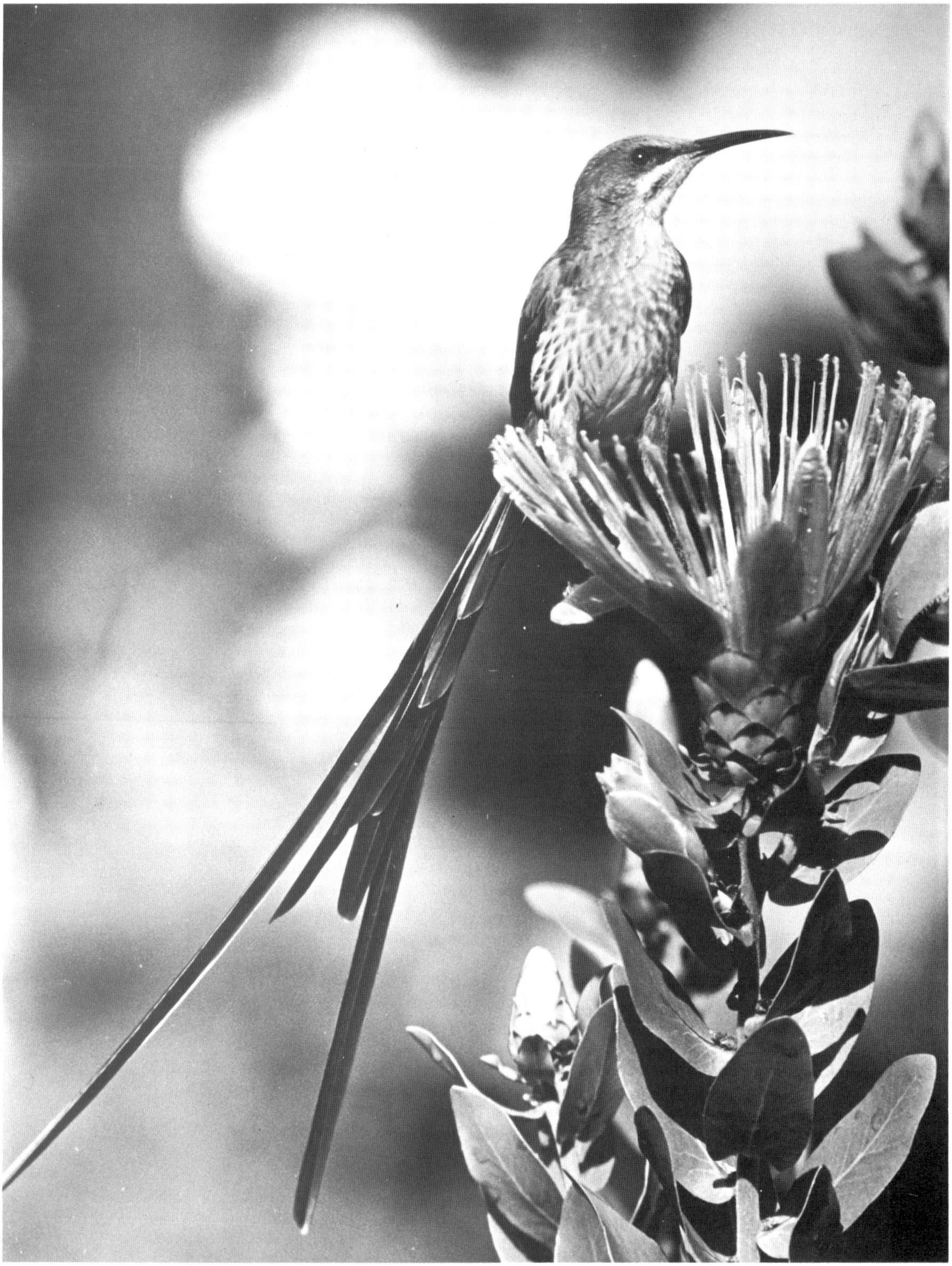

Orangebreasted Sunbird (*Nectarinia violacea*)

Seeds of the Protea Family

Seeds of the protea family vary considerably, as may be seen in the design above, which was made by grouping various types of seeds. These are numbered in the photograph for those who are interested in identifying them.

1. *Protea neriifolia*
2. *Leucadendron daphnoides*
3. *Leucadendron album*
4. *Protea compacta*
5. *Serruria pedunculata*
6. *Mimetes cucullatus*
7. *Leucadendron gandogeri*
8. *Protea repens*
9. *Protea magnifica*
10. *Aulax umbellata*
11. *Protea cynaroides*
12. *Protea speciosa* (infertile)
13. *Leucadendron argenteum*
14. *Leucadendron platyspermum*
15. *Leucadendron rubrum*
16. **Protea burchellii**
17. *Leucospermum reflexum*

PROTEAS AS A COMMERCIAL CUT FLOWER CROP

The value of Proteas as cut flowers has led to the development of an increasing market for them among the florists of the world, notably in Europe. A great marketing effort by specialist importers, plus improved air-freight facilities since the early 1970s, has made it possible for South African growers to send increasing quantities to Amsterdam, Düsseldorf, Frankfurt and Zürich. A remarkable amount of proteaceous material, particularly of types of *Leucadendron*, is exported from the Cape under the name of 'Cape Greens', and is very popular in Europe. An intercontinental market has developed since 1975, when Proteas came to be grown commercially in areas such as California, Australia, New Zealand, Hawaii, on the island of Maui, and Israel. Although these countries cannot compete with the quantities of Proteas that emanate from their natural habitat and plantations in the Cape, they are of some value to European markets, extending the availability of various species, because of the difference in seasons between the northern and southern hemispheres and the diverse environments which produce slightly different flowering periods.

The delightful variability in seed-grown Proteas is not always favoured by florists, who need to plan their arrangements in advance. The development of new cultivars that are grown from cuttings and planted in the mass has been an important step in the cut flower market, enabling the florist to rely upon consistency in flower sizes, shapes and colours. Some Cape farmers, who have been pioneers in the field of producing for the Protea flower trade, have experimented with cuttings from especially beautiful natural specimens in the wild and increased them, giving them special names to popularise and identify them. They have succeeded in developing these specialised cultivars, which are proving to be popular export flowers. Research in the development of superior cultivars is taking place at government horticultural research stations in the Cape, with the aim of distributing them to growers as soon as they prove successful. This research will, no doubt, benefit farmers who are endeavouring to meet the demands of an ever-increasing market, both at home and abroad. Similar efforts are being made in Hawaii.

Problems of transport have led to experimentation in picking and packing methods for Proteas, so as to ensure that the flowers may arrive at their destination in good condition. Commercial growers have found that the greatest enemies of Proteas *en route* to distant markets are high temperatures and dehydration. Although Proteas are sturdy and good travellers, specialised treatment when picking and packing will assist the flowers to arrive as fresh as possible.

Proteas should be placed in clean water as soon as possible after picking, to stand in the water for at least one hour. To precool Proteas in a cold room for several hours after that is advisable, if possible. Some growers place them in a chemical sugar solution, but this is not done in South Africa. The lower leaves should be stripped off as their presence serves to accelerate transpiration and, as they will ultimately be removed by the florist before being arranged, it is both practical and economical to do so at harvesting time, especially as it reduces weight for air travel.

The actual packing of Proteas for the export trade is no more complicated than the packing of any fresh flowers which must travel long distances. Much depends on the size and weight of the type of Protea, which should be packed in sturdy boxes. Fairly tight packing in order to prevent movement is the rule for packing any flowers for travelling.

Experimentation into the feasibility of sending the flowers abroad by sea in containers, as well as into fumigating methods to destroy insects and the problem of foliage turning black during transportation, is taking place continuously, both in South Africa and Hawaii. Careful growers, who wish to perfect the quality of their blooms, check their condition on arrival at their destination and experiment constantly on how to improve their methods of picking and packing. No doubt these problems will continue to be a subject of research at all times in the future because of the steady growth in demand for Proteas by all florists, especially as the export of Protea flowers is a huge and growing enterprise.

The life of a flower which is cut freshly from a bush in the garden and then placed immediately in the vase varies from ten days to a month, according to type. Some look fresh for up to a month, such as *Leucospermum cordifolium*, while *L. reflexum* generally starts to reflex after a few days and looks untidy after a week. It is difficult to decide when a Protea ceases to look fresh and when it is slowly transformed into dried flower material, for much depends on the taste of the person who owns the flower arrangement. On the whole, Proteas look really fresh for about ten days, but have the added advantage that they fade slowly into soft tapestry colours which have an attraction of their own and are certainly longer-lasting than most fresh flowers.

THE PHOTOGRAPHS

COLOUR PHOTOGRAPHS BY SIMA ELIOVSON

Page
207 *Aulax umbellata*
211 *Diastella divaricata*
109 *Protea compacta*
141 *P. cordata*
 93 *P. effusa*
115 *P. eximia*
121 *P. lacticolor*
111 *P. lepidocarpodendron*
107 *P. longifolia*
129 *P. coronata*
125 *P. pudens*
123 *P. mundii*
127 *P. nana*
119 *P. burchellii*
 87 *P. rubropilosa*
139 *P. scolopendriifolia*
131 *P. scolymocephala*
113 *P. susannae*
 87 *P. laetans*
 89 Proteas from cuttings, air-layers
155 *Leucospermum catherinae*
151 *L. conocarpodendrum*
165 *L. glabrum*
163 *L. lineare*
149 *L. reflexum*
167 *L. truncatum*

Page
165 *Leucospermum vestitum*
187 *Leucadendron album*
177 *L. arcuatum*
183 *L. coniferum*
181 *L. discolor*
181 *L. eucalyptifolium*
189 *L. galpinii*
175 *L. globosum*
189 *L. linifolium*
183 *L. muirii*
185 *L. platyspermum*
185 *L. rubrum*
179 *L. salignum*
177 *L. sessile*
187 *L. uliginosum*
179 *L. xanthoconus*
191 *Mimetes cucullatus*
195 *M. hirtus*
203 *Paranomus reflexus*
199 *Serruria barbigera*
197 *S. florida*
199 *S. burmanii*
201 *S. pedunculata*
211 *Spatalla curvifolia*
217 *Telopea speciosissima*

COLOUR PHOTOGRAPHS BY EZRA ELIOVSON

Page
133 *Protea acaulos*
103 *P. subulifolia*
145 *P. aspera*
101 *P. amplexicaulis*
 73 *P. nitida*
 67 *P. magnifica*
 83 *P. caffra*
 95 *P. acuminata*
 65 *P. cynaroides*
 77 *P. grandiceps*
 89 *P.* hybrid
 81 *P. laurifolia*
105 *P. aurea*
 79 *P. neriifolia*
 71 *P. obtusifolia*
 69 *P. repens*
 85 *P. roupelliae*

Page
 75 *Protea speciosa*
 99 *P. sulphurea*
 ii *Protea* group-frontispiece
173 *Leucospermum bolusii*
147 *L. cordifolium*
161 *L. gerrardii*
171 *L. mundii*
171 *L. oleifolium*
159 *L. prostratum*
157 *L. praecox*
173 *L. rodolentum*
153 *L. tottum*
175 *Leucadendron argenteum*
209 *Orothamnus zeyheri*
203 *Paranomus spicatus*
201 *Serruria aemula*
201 *S.* hybrid

COLOUR PHOTOGRAPHS BY PROFESSOR H. B. RYCROFT
by courtesy of the National Botanic Gardens, Kirstenbosch

Page

205 *Brabejum stellatifolium*
143 *Protea cryophila*
139 *P. humiflora*
145 *P. restionifolia*

97 *Protea pendula*
117 *P. stokoei*
135 *P. witzenbergiana*
193 *Mimetes argenteus*

COLOUR PHOTOGRAPHS BY DOREEN JEFFES

169 *Leucospermum grandiflorum*
213 *Faurea saligna*

COLOUR PHOTOGRAPH BY DR. P. P. DU TOIT

93 *Protea pityphylla*

COLOUR PHOTOGRAPH BY DR. P. J. JOUBERT

137 *Protea venusta*

COLOUR PHOTOGRAPH BY J. LEIBOWITZ

91 *Protea aristata*

PHOTOGRAPHS BY PROFESSOR C. J. UYS

219 Cape Sugarbird
220 Orangebreasted Sunbird
    (by permission from *Halcyon Days*)

PHOTOGRAPH BY VIC. JACOBS

221 *Protea* seeds, arranged by Sima Eliovson

MONOGRAPHS PUBLISHED ON THE GENERA OF THE SOUTH AFRICAN PROTEACEAE

"*Taxonomic Studies on Leucospermum* R. Br." by J.P. Rourke, supplementary Vol. No. 8 of the *Journal of South African Botany*, April, 1972.

"*Taxonomic Studies on Sorocephalus and Spatalla*" by J.P. Rourke, supplementary Vol. No. 7 of the *Journal of South African Botany*, September, 1969.

Revision of the genus *Paranomus* by Dr. M.R. Levyns, published as *Contributions from the Bolus Herbarium* No. 2, 1970.

Revision of the genus *Leucadendron* by Dr. Ion Williams, published as a further part of *Contributions from the Bolus Herbarium*.

The Proteas of Southern Africa by J. P. Rourke.

INDEX

Scientific names appear in *italics*. Numbers in **heavy** type indicate more detailed descriptions.

Aardroos, 43, **124**
Acacia, 7
Acidity of Soil, 21, 22
Adenanthos, 9
Agastachys, 9
Ageratum, 49
Ajuga reptans, 49
Alkalinity of Soil, 21, 22, 23
Almond, Wild, **204**
Alyssum, 49
Amandelhout, **204**
Antirrhinium, 4
Arctotheca calendula, 49
Arctotis, 5
Artichoke, 64
Arum Lily, White, xviii
Aulax, 9, **206**
   cancellata, 45, 47, 54, **206**
   cneorifolia, **206**
   pinifolia, **206**
   umbellata, 23, 44, 45, 47, 54, **206**

Banksia, 9, 214, 215
   ericifolia, **214**
Beech, European, 212
Bergroos, **126**
Berzelia lanuginosa, ii
Blousuikerbos, **78**
Blushing Bride, xx, 16, 24, 38, 44, 58, 61, 86, **196**
Boekenhout, Transvaal or Red, **212**
Bossiestroop, xviii
Bottlebrush,
   Green, 44, **202**
   Red, 44, **190**
   Red-and-Yellow, **194**
   Silver-leaved, **192**
Bougainvillea, MacLean, 61
Brabejum, 9, 13
   stellatifolium, 53, **204**

Calendula, 5
Cerastium, 49
Chilean Fire-Bush, **215**
Cluster Pine, 76
Cone-Bush, **215**
Conospermum, 9, **215**
Cotula barbata, 49
Cripple-wood, **150**
Cynara, 64

Diastella, 9, **210**
   divaricata, 54, **210**
   ericacifolia, **210**
   serpyllifolia, **210**
Dimorphic foliage, 184, 188, 202
Dimorphotheca, 5
Dorotheanthus, 49
Dryandra, 9, **215**

Embothrium, 9, **215**
   coccineum, **215**

Faurea, 9, 52, **212**
   saligna, **212**
   speciosa, **212**
Feather-dusters, **206**
Fire-Tree, Chilean, **215**
Firewheel Tree, **214**
Fynbos, 7, 194, 210

Gazanias, tufted, 49
Geebung, 216
Geelbos, 44, **178**
Gerbera jamesonii, 49
Gevuina, 9
Gladiolus alatus, 5
Gold-Tips, 5, 44, **178**
   Ceres, **176**
   Flame, **180**
Gousblom, 5
Grey leaves, 186
Grevillea, 9, **214**
   banksii, **214**
   robusta, **214**
   rosmarinifolia, **214**
Grevillea, Rosemary, **214**
Grevillea, Scarlet, **214**
Ground-Rose, **124**

Hakea, 7, 10, **215**
   acicularis, 215
   saligna, 215
   sericea, 215
Hicksbeachia, 10
Hypericum, 61

Ipomoea tuberosa, 57
Isopogon, 10, **215**

Kalkoentjie, 5
Knightia, 9
Knobkerrie Bush, 5, **184**
Kreupelhout, 44, **150**

Lambertia, 9
Lampranthus, 49
Lavandula, 4
Lavender, 4
Leucadendron, 7, 9, 12, 34, 35, 40, 47
  52, 60, 176, 186, 202, 206
  adscendens, 178
  album, 45, 46, **186**
  aemulum, 188
  arcuatum, 27, 45, **176**
  argenteum, 27, 30, 43, 50, 51, 52,
    54, **174**, 186
  aurantiacum, 27, 54, 55, 186
  comosum, 53, **188**
  cinereum, 53, **188**
  coniferum, 53, **182**
  crassifolium, 176
  daphnoides, 45, 53, **176**
  decorum, 176
  discolor, 43, 45, 47, 52, 88, **180**
  eucalyptifolium, 45, 53, **180**
  galpinii, **188**
  gandogeri, 53, 55, **176**
  globosum, **174**
  grandiflorum, 176
  guthrieae, 176
  lanigerum, 53, **176**
  laureolum, 52, **176**
  linifolium, 54, **188**
  microcephalum, 54, **176**
  muirii, 23, 53, **182**
  olens, **174**
  platyspermum, 5, 34, 53, **184**
  plumosum, 184
  pubescens, 54, **186**
  repens, 68
  rubrum, 27, 34, 35, 53, **184**, 186
  sabulosum, 182
  salicifolium, 178
  salignum, 53, 55, **178**
  sessile, 54, **176**
  sericocephalum, 186
  stokoei, 176
  strictum, 19, 53, **178**
  tinctum, 5, 27, 44, 53, **176**
  tortum, 188

Leucadendron—Cont.
  uliginosum, 19, 23, 54, **186**
  venosum, 176
  xanthoconus, **178**
Leucadendron,
  Galpin's, **188**
  Large Pink-Coned, **182**
  Silver-Coned, **182**
Leucospermum, xx, 7, 9, 12, 27, 29,
  30, 34, 35, 40, 60, 86
  album, 172
  attenuatum, 156
  bolusii, 45, 46, 53, 146, **172**
  candicans, 44, 45, 53, 54, **172**
  catherinae, 27, 43, 45, 46, 47,
    52, **154**
  ceresis, 158
  conocarpodendron, 44, 45, 46, 47, 52, **150**
  conocarpum, 150
  cordifolium, 23, 39, 43, 45, 46, 47,
    50, 51, 53, 57, 58, 61, **146**,
    152, 154, 162, 164
  crinitum, 35, 44, 45, 46, 47, 53,
    55, 156, 170
  cuneiforme, 23, 43, 45, 46, 53, **156**, 160
  ellipticum, 164
  gerrardii, 45, 53, 55, **160**
  glabrum, **164**
  grandiflorum, 27, 44, 45, 46, 53,
    156, **168**
  hypophyllocarpodendron, 55, **158**
  hypophyllum, 158
  incisum, 43, 45, 46, 47, 53, 164
  innovans, **160**
  lineare, 38, 43, 45, 46, 51, 53, 54,
    55, **162**
  medium, 164
  muirii, 44, 45, 46, 53, 55, **166**
  mundii, **170**
  nutans, 146
  oleifolium, 44, 45, 53, 55, 156, **170**
  praecox, 88, **156**, 170
  prostratum, 44, 45, 50, 55, 156, **158**
  reflexum, 29, 38, 43, 45, 46, 50, 52,
    54, 62, 88, 146, **148**
  saxosum, **160**
  rodolentum, 44, **172**
  spathulatum, **158**
  tottum, 43, 44, 47, 53, 88, **152**, 158
  truncatum, **166**
  vestitum, 45, 47, 53, 146, **164**
Lomatia, 9

Macadamia, 9, 40, **215**
  ternifolia, **215**
Macadamia Nut, **215**
Macchia, 7, 194, 210
Marsh Rose, 17, 42, **208**
Mesembryanthemum, 49
Mimetes, 7, 9, 40, 190
  argenteus, 53, 54, 55, 190, **192**
  capitulatus, **190**
  cucullatus, 44, 45, 47, 53, 88, **190**
  hartogi, **190**
  hirtus, 45, 47, 53, 190, **194**
  hottentoticus, **190**, 192
  integrus, **192**
  lyrigera, 44, 45, 47, 53, 55, 190
Mountain Rose, 23, 43, **126**
  Long-leaved, **92**

Nectarinia violacea, **218**, 220
Nerium oleander, 78

Oleander, 78
Orothamnus, 9, 52
  zeyheri, 17, 42, **208**

Paranomus, 9, 12
  crithmifolius, 53, 54, **202**
  reflexus, 44, 45, 46, 47, 53, **202**
  sceptrum-gustavianum, 53, **202**
  spicatus, 44, 45, 47, 53, 54, **202**, 216
Perdebos, 44
Persoonia, 9, **216**
Petrophila, 9, **216**
Phlox, Alpine, 49
Pincushion,
  Catherine's, 43, **154**
  Creeping, **158**
  Dwarf, **160**
  Fire-wheel, 43, **152**
  Large Tufted, **156**
  Narrow-leaf, 43, **162**
  Nodding, 43, **146**
  Notched, **164**
  Rainbow, 44, **168**
  Rocket, 43, **148**
  Spreading, **152**
  Tufted, 44, **170**
  Upright, 43, **164**
  White, **172**
Pincushion Flowers, xx, 27, 29, 30, 58, 61
Pinus pinaster, 7, 76
Portulaca, 49

Promerops cafer, **218**
Protea, 3, 4, 6, 9, 30, 40, 196
  abyssinica, 82
  acaulos, 45, 54, 60, 102, **132**, 198
  acerosa, 102
  acuminata, 27, 45, 46, 53, 55, 60, **94**, 98, 130, 142
  amplexicaulis, 27, 41, 50, 54, 55, **100**, 102, 136
  arborea, 72
  aristata, 16, 27, 43, 45, 47, 50, 53, 54, 55, **90**, 102
  aspera, 55, 102, **144**
  aurea, 45, 47, 52, 59, **104**, 218
  barbigera, 53, 66
  burchellii, 38, 42, 44, 45, 47, 53, 59, 61, 62, **118**
  caffra, xviii, 24, 44, 47, 52, **82**
  canaliculata, 27, 55, **94**
  cedromontana, 94
  compacta, 17, 42, 44, 46, 52, 59 **108**
  comptonii, **86**
  cordata, 41, 54, 102, **140**
  coronata, xx, 43, 45, 46, 52, **128**
  cryophila, 27, 41, 54, 92, 102, **142**
  cynaroides, xix, 5, 17, 22, 23, 35, 38, 42, 44, 45, 46, 47, 50, 53, 55, 58, 59, 60, **64**, 88
  echinulata, 144
  effusa, 27, 45, 53, 55, 60, **92**
  eximia, 27, 38, 45, 46, 47, 53, 84, 88, **114**
  gaguedi, **82**
  grandiceps, 5, 27, 42, 45, 46, 47, 53, **76**, 88, 98
  grandiflora, 72
  harmeri, 94
  hirta, 59, 82
  humiflora, 27, 54, 55, 102, **138**
  Hybrids, **88**
  incompta, 128
  lacticolor, 43, 44, 46, 52, 59, **120**, 122
  laetans, 45, **86**
  lanceolata, 59, **122**
  latifolia, 27, 53, 84, 114
  laurifolia, 42, 44, 46, 52, 78, **80**, 106
  lepidocarpodendron, 43, 45, 46, 53, **110**
  longiflora, 104
    var. *minor*, 124
  longifolia, 43, 45, 46, 53, **106**

228

Protea—Cont.
  lorea, 55, 102, **144**
  lorifolia, 27, 53, **80**, 142
  macrocephala, 128
  macrophylla, 27, 53, 59, 80
  magnifica, 5, 17, 27, 34, 35, 42, 44, 45, 46, 50, 53, 58, 59, 60, **66**, 88, 100, 106
  marginata, 52, 80
  marlothii, 92
  mellifera, xviii, 4, 5, 16, 53, 68
  minor, 124
  multibracteata, 82
  mundii, 44, 46, 47, 53, 120, **122**
  nana, xix, 6, 24, 43, 45, 46, 50, 53, 54, 55, 60, 61, 92, 102, **126**, 134
  neriifolia, xx, 11, 23, 35, 42, 44, 46, 53, 58, 62, **78**, 80, 88
  nitida, 45, 46, 52, **72**
  obtusifolia, 22, 23, 38, 42, 45, 46, 47, 53, 59, **70**, 78, 88
  odorata, xix, 54, **124**
  parvula, **120**
  patersonii, 74
  pendula, 55, **96**, 134
  pityphylla, 27, 43, 45, 54, **92**, 96, 102, 126, 134
  pudens, 23, 43, 45, 46, 54, 55, **124**
  pulchella, 118
  pulchra, 118
  punctata, 27, 44, 46, 53, 88, **120**
  repens, xviii, xx, 4, 16, 23, 27, 34, 35, 39, 42, 44, 45, 46, 47, 53, 60, **68**, 70, 88, 118
  restionifolia, 55, **144**
  rhodantha, 82
  rosacea, 126
  roupelliae, 24, 44, 47, 53, 82, **84**, 86
  rubropilosa, 27, 47, 53, **86**
  scabra, 55, 102, **144**
  scolopendriifolia, **138**
  scolopendrium, 138
  scolymocephala, 45, 46, 47, 53, **130**
  simplex, **82**
  speciosa, xx, 5, 17, 42, 44, 46, 47, 53, 58, 59, **74**, 116
  stokoei, 15, 43, 44, 46, 53, 74, **116**
  subpulchella, 118
  subulifolia, 23, 54, **102**, 140
  subvestita, **120**
  sulphurea, 55, **98**, 136
  susannae, 23, 43, 44, 46, 47, 53, **112**
  venusta, 27, 55, **136**

Protea—Cont.
  welwitschii, 59, **82**
  witzenbergiana, 27, 54, 55, 96, 102, **134**
Protea,
  Apple Green, **128**
  Baby, 43, **120**
  Black-bearded, **110**
  Blyde, **86**
  Bot River, 42, **108**
  Brown-bearded, 17, 42, **74**
  Cedarberg, **94**
  Drakensberg, **84**
  Dwarf Green, **132**
  Fringed, 42, **80**
  Giant, **64**
  Giant Woolly-beard, xx, 17, 42, **66**
  Gleaming, 42, **118**
  Grass-leaved, **144**
  Green,
    Apple Green, 43, **128**
    Dwarf Green, **132**
    Small Green, **130**
  Ground, **138**
  Heart-leaf, **140**
  Highveld, **82**
  Ivy-leaved, **100**
  King, 5, 42, **64**
  Laurel-leaved, **80**
  Ladismith, 43, **90**
  Long-Bud, **104**
  Long-leaved, 43, **106**
  Mund's, **122**
  Needle-leaved, **102**
  Nodding, **96**
  Oleander-leaved, 42, **78**, 80
  Oval-leaved, **76**
  Peach, 42, **76**
  Pine-leaved, 43
  Prince, 5, **74**
  Princess, 5, **76**
  Queen, 5, **66**
  Ray-flowered, **114**
  Roupell's, **84**
  Small Green, **130**
  Snow, **142**
  Stem-clasped, **100**
  Stokoe's, 43, **116**
  Sulphur-coloured, **98**
  Susan's, 43, **112**
  Swartberg, **136**
  Velvet, **86**
  Witzenberg, **134**

229

Proteaceae, xx, 3, 6, 12, 39
Proteas,
    Frost-resistant, 27
    Lime-tolerant, 23

Queensland Nut, 215

Rooistompie, **190**
Rose Cockade, 5, 44, **176**
Rosemary Grevillea, 214
Roupala, 9

Scarlet Grevillea, 214
Seeds of Protea family, **221**
Serruria, xx, 9, 12
    adscendens, **198**
    aemula, 46, 86, **200**
    artemisiaefolia, 200
    barbigera, 45, 53, 54, 55, **198**
    burmanii, **198**
    florida, xx, 16, 24, 38, 44, 45, 46,
        50, 53, 54, 55, 58, 61, 86, **196**,
        198, 200
    knightii, 198
    pedunculata, 46, 53, 54, **200**
    pinnata, **198**
Serruria,
    Grey, **200**
    Silky, **198**
Silver Bush, **186**
Silver or Grey Leaves, **186**
Silver Tree, 27, 28, 30, 43, **174**, **186**, 192
Skaamblom, **126**
Smoke-bush, **215**
Snapdragon, 4
Sneeublom, 92, **142**
Snow Flower, **142**
Soldaat, **190**

Sorocephalus, 9, **210**
Spatalla, 9, **210**
    curvifolia, **210**
Spatallopsis, 9, 210
Spider Bush, **198**
Spinnekopbos, **198**
Stenocarpus, 10, **214**
    sinuatus, 214
Sugarbush, xviii, xx, 5, 16, **68**
    Bredasdorp, **70**
    True, 42, **68**
Sugarbird,
    Cape, **218**
    Gurney's, **218**
Suikerbos, xviii, xx, 5, 16
    Reuse, **64**
    Ware, 42, **68**
Suikerbos stroop, xviii
Sunbird, Orangebreasted, 218, **220**
Sunshine Bush, 43, **180**

Telopea, 10, **214**
    speciosissima, **214**, 216, 217
Tolbos, **184**
Top-bush, **184**
Trots van Fransch Hoek, **196**

Vegetative Reproduction, 86

Waboom, **72**
Waratah, **214**, 216, 217
Wild Almond, **204**
Wilde Amandel, **204**
Witteboom, **174**
Woodrose, **57**
Woody-Pear, **216**

Xylomelum, 10, **216**

Zantedeschia aethiopica, xviii

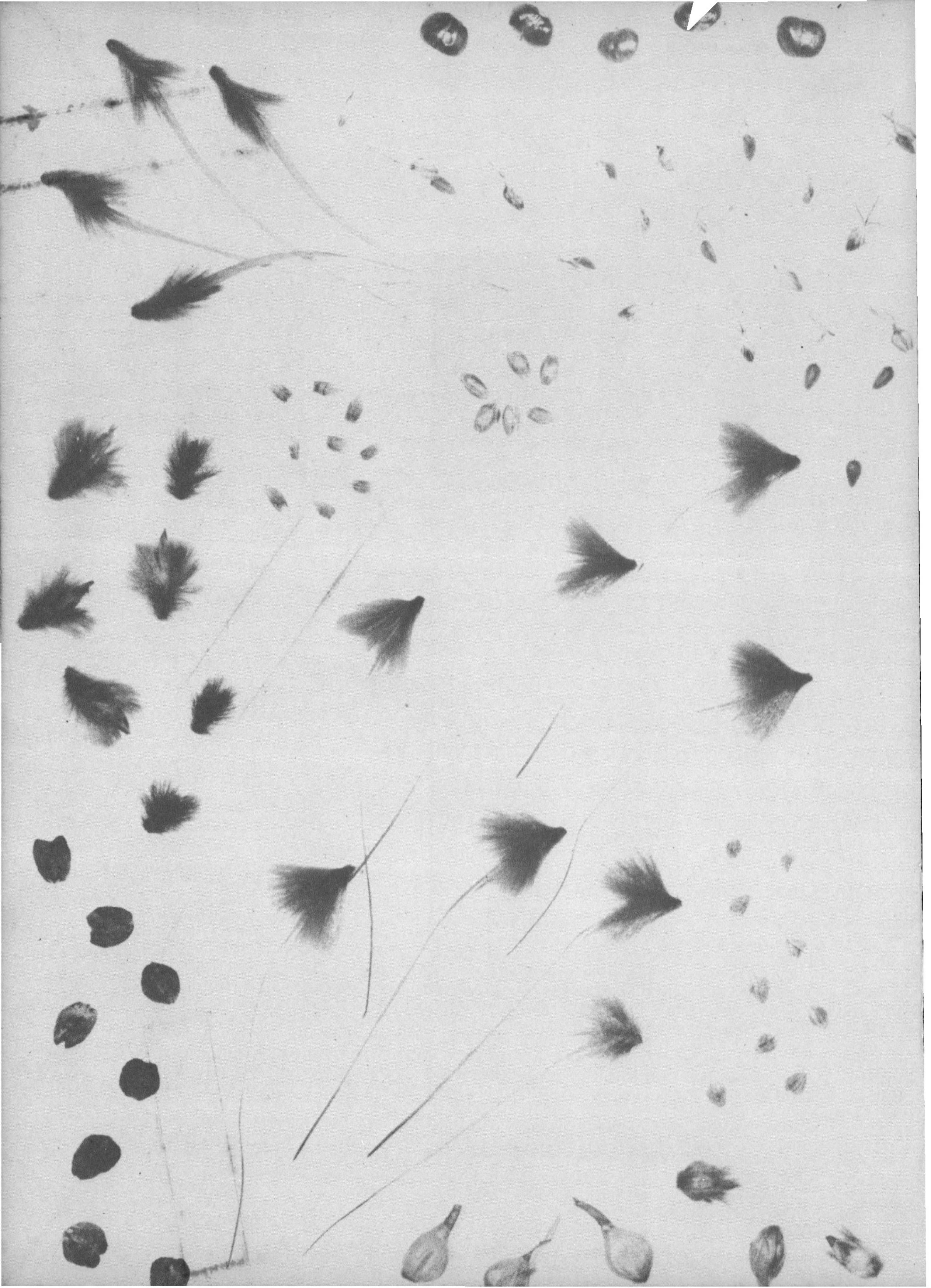